T0203025

Process Integration Approaches to Planning Carbon Management Networks

Green Chemistry and Chemical Engineering

Series Editor:
Sunggyu Lee
Ohio University, Athens, Ohio, USA

Process Integration Approaches to Planning Carbon Management Networks

Dominic C. Y. Foo
and
Raymond R. Tan

CRC Press
Taylor & Francis Group
Boca Raton London New York

CRC Press is an imprint of the
Taylor & Francis Group, an **informa** business

LINGO® is a registered trademark of Lindo Systems, Inc.

First edition published 2020
by CRC Press
6000 Broken Sound Parkway NW, Suite 300, Boca Raton, FL 33487-2742

and by CRC Press
2 Park Square, Milton Park, Abingdon, Oxon, OX14 4RN

First issued in paperback 2021

© 2020 Taylor & Francis Group, LLC

CRC Press is an imprint of Taylor & Francis Group, an Informa business

Publisher's Note
The publisher has gone to great lengths to ensure the quality of this reprint but points out that some imperfections in the original copies may be apparent.

Typeset in Palatino
by codeMantra

ISBN 13: 978-1-03-224281-1 (pbk)
ISBN 13: 978-0-8153-9092-3 (hbk)

DOI: 10.1201/9781351170888

Dominic C. Y. Foo would like to dedicate this book to his wife

Cecilia, and their daughters, Irene, Jessica, and Helena.

Raymond R. Tan dedicates this book to his daughters, Denise and Dana.

The kids serve as constant reminders that the current generation must leave

behind a better world for the future inhabitants of our common home.

Contents

Part B   Applications of Carbon Management Networks

Preface

Despite politically driven climate change denial in some parts of the world, there is now an overwhelming scientific consensus that anthropogenic greenhouse gas emissions are causing global warming. The warming appears to have accelerated, and in recent years, we have witnessed inauspicious records being set as evidence of climate change accumulates. Atmospheric carbon dioxide (CO_2) concentration now exceeds 400 ppm, which is a level not seen in human history, and in the past few years recorded mean global temperature has climbed noticeably. Fortunately, the gravity of the problem has gained recognition in the public eye and among policy makers, as indicated for example by climate action being listed as one of the United Nations Sustainable Development Goals (UN SDG). Many countries have also committed to voluntary cuts in their emissions, most recently via the Paris Agreement (during the 21st session of the Conference of Parties – COP21). The challenge still remains as to how to reduce carbon emissions where they actually occur, for example in the exhaust gases of power plants, industrial sites, motor vehicles, and aircraft, among others.

We initially developed *carbon emissions pinch analysis* (CEPA) in late 2005 as a novel extension of *pinch analysis* techniques developed in the 1970s for heat recovery and energy conservation in industrial plants. The latter is a sub-discipline of chemical engineering known as *process integration*. Our paper "Pinch Analysis Approach to Carbon-Constrained Energy Sector Planning" was eventually published in early 2007 in the journal *Energy*. In this seminal work, we showed how pinch analysis concepts and methodology can be applied to the problem of allocating energy sources to different demands while accounting for carbon emissions constraints. Our initial graphical approach attracted the attention of research groups in Ireland and New Zealand within a year or so of its publication. CEPA has since developed as a small but growing branch of pinch analysis and process integration literature. To date, the original paper has been cited more than 200 times (Scopus database); different research teams, including our own, have developed variants and applied these to different problem contexts in six different continents. In addition to research articles, the current CEPA literature also includes review papers as well as chapters in handbooks and encyclopedias. We recently coined the phrase "carbon management network" or CMN as a generic term for systems designed via CEPA or its variants for the purpose of managing carbon emissions. This book marks a natural step in the maturation of CEPA as a sub-area in process integration.

This hybrid book is intended to serve a dual purpose. We have written the first five chapters as a textbook-style tutorial for beginners seeking to learn CEPA methodology. The discussion is written at the level that is appropriate for advanced undergraduate and postgraduate students as well as professionals in strong quantitative disciplines such as Engineering, Economics, or Physics. As long as the reader has a rudimentary understanding of the physical nature of energy systems and climate change, the depth of discussion should be accessible. We also provide a description of alternative approaches to CEPA to suit reader preferences. In principle, the graphical techniques can be roughly implemented with pen, ruler, and paper, but the use of Excel spreadsheets and dedicated commercial optimization software such as LINGO are also included. Chapter 1 first provides an introduction to the general problem of carbon-constrained energy planning. Then, Chapter 2 discusses the graphical approach developed in our 2007 paper, which is used to determine the minimum amount of zero- or low-carbon energy resource needed to meet emissions cuts in a given system. The chapter also shows how the same methodology can be modified to apply for cases where sustainability metrics other than carbon footprint are of interest. Chapter 3 then covers algebraic targeting and *automated targeting model* (ATM), which are variants of the basic CEPA approach. Although mathematically equivalent, these extensions can be readily implemented using various software, and can also be modified more readily to handle case-specific issues. Chapter 4 deals with the application of CEPA extensions for planning of CO_2 *capture and storage* (CCS) systems. In particular, planning CCS must account for energy balance problem of how to compensate for capacity reduction resulting from CO_2 capture retrofits in power plants, and the mass balance problem of allocating CO_2 among sources and sinks in a CMN. The main concept for CEPA approaches in Chapters 2–4 are meant to set performance targets for the system, following with the concept of conventional pinch analysis techniques. Finally, Chapter 5 discusses an alternative approach based on *superstructure* models, which allow the user to bypass the equivalent stepwise pinch analysis approach. The procedural efficiency of this alternative comes at the expense of loss of transparency. Problems solved in earlier chapters are revisited using superstructure models to illustrate the equivalence of these two complementary methodologies. In addition to each chapter's core content, we also provide suggestions for the advanced reader to explore further.

The second half of this book is an edited compilation of contributed chapters from different research groups. These chapters provide a means to introduce a more advanced reader, perhaps intent on doing further CEPA research himself/herself, to current developments and applications. These chapters show how CEPA can be applied for different problems in different countries, each with unique features and at a particular stage of economic developmental. Chapter 6 by Jia, Li, and Wang discusses the use of CEPA in China, the world's largest country in terms of both population and greenhouse gas emissions. This chapter also reviews some of the interesting local CEPA literature in China, which is indicative of interest among Chinese scholars in planning low-carbon growth. In Chapter 7, Jain and Bandyopadhyay apply CEPA to energy planning in India, which in many ways is similar to China, and is projected in the near future to have the largest population in the world. Chapter 8 by Lim and Tarun, on the other hand, discusses the use of ATM in the United Arab Emirates (UAE), a relatively small country but one with very high per capita carbon emissions. In addition, due to its arid climate, the water and energy issue in UAE is particularly pronounced. Then, in Chapter 9, Andiappan, Ng, and Foo illustrate the combination use of CEPA and superstructural model at the supply chain level in Malaysia. This work shows the potential for use of abundant biomass resources from agro-industrial systems as a low-carbon energy source. Finally, Chapter 10 by Lee describes carbon-constrained energy planning in Taiwan, an industrialized but small and resource-constrained island. Unlike the previous four chapters, Lee makes use of a mathematical programming approach. Although these chapters were written to be stand-alone contributions, the reader may easily cross reference any unclear concepts with the CEPA tutorials in the first half of the book. Supplementary files such as spreadsheet and mathematical models are also available in the book's supporting website (www.crcpress.com/9780815390923) for the ease of use for readers (please look for logo ⌨ in the book).

The ten chapters of this book provide an entry point for any reader to explore CEPA and CMN, which represents our modest scientific contribution to the looming problem of climate change.

Dominic C. Y. Foo and Raymond R. Tan
January 2020

LINGO® is a registered trademark of Lindo Systems, Inc. for Product Information, please contact:

Lindo Systems
1415 North Dayton Street
Chicago, IL, 60642 USA
Tel: (800) 441-BEST(2378), (312)988-7422, Support: (312)988-9421
Fax: (312)988-9065
E-mail: info@lindo.com
Web: www.lindo.com

Authors

Dominic C. Y. Foo is a professor of process design and integration at the University of Nottingham Malaysia. He is a fellow of the Institution of Chemical Engineers, a fellow of the Academy of Science Malaysia, a chartered engineer with the UK Engineering Council, and a professional engineer with the Board of Engineers Malaysia. He works on process integration for resource conservation and CO_2 reduction, with more than 400 published works. Prof. Foo is the co-editor-in-chief for *Process Integration and Optimization for Sustainability*, subject editor for *Process Safety & Environmental Protection*, and an editorial board member for several other renowned journals.

Raymond R. Tan is a professor of chemical engineering and university fellow at De La Salle University, Philippines. He is also a member of the National Academy of Science and Technology of the Philippines. His main areas of research are process systems engineering and process integration, where he has over 300 published works. Prof. Tan received his BS and MS degrees in chemical engineering and PhD in mechanical engineering from De La Salle University. He is also co-editor-in-chief of *Process Integration and Optimization for Sustainability*, subject editor of *Sustainable Production and Consumption*, and an editorial board member of *Clean Technologies and Environmental Policy*.

Part A

Basic Methodology for Carbon Management Networks

1

Global Energy and Climate Landscape

1.1 Introduction

There is a strong global scientific consensus that emissions of greenhouse gases (GHGs), such as carbon dioxide (CO_2), methane (CH_4), and nitrous oxide (N_2O), from human activities are driving climate change (IPCC, 2018). Atmospheric concentration of CO_2 now exceeds 400 ppm, which is well above the levels that prevailed in pre-industrial times. Emission levels have also been steadily increasing due to economic and demographic trends. Population growth coupled with rising standards of living have led to increased emissions resulting from fossil energy use for provision of electricity, industrial goods, and transportation services, while shifts in dietary preferences have also led to increased GHG emissions from agriculture. Business-as-usual (BAU) GHG emissions trends are expected to cause disruptive changes in the Earth's climate system, with mean atmospheric temperature rising by more than 2°C by the end of the century. On the other hand, sufficiently deep cuts in global GHG emissions can limit temperature rise to a relatively safe and manageable level of 1.5°C. However, due to the complex dynamics of atmospheric carbon and energy balances, the effects of emissions reductions are also highly time-dependent. Dramatic reductions will need to be implemented in the short and medium terms, so as to reduce net CO_2 emissions to zero by approximately mid-century (IPCC, 2018). However, even if such deep cuts are achieved, the radiative forcing of carbon stock already in the atmosphere will continue to cause mean temperature to rise in subsequent decades.

There are numerous potential adverse risks that can result even from a seemingly modest increase in mean temperature in the range of 1.5°C–2°C (IPCC, 2018). For example, sea level is expected to rise through expansion of seawater and melting of polar ice in the Arctic and Antarctic. Such sea level changes can then result in potential displacement of people from many densely populated, low-lying coastal cities and communities. Furthermore, actual temperature changes in specific geographical locales can deviate from the global mean level. Significant risks can result from the complexity of climate and weather systems, causing temperature deviations away from historical levels to which inhabitants are accustomed. Temperature anomalies may thus be positive (e.g., heat waves) or, counterintuitively, negative (e.g., abnormally cold conditions). Other risks pertain to changes in precipitation patterns, which can have major implications on major population centers that rely on rain to sustain water supply for agriculture, domestic use, electricity generation, and other economic activities. Increased thermal energy in warmer seas can also result in increased frequency of extreme weather events, thus leading to more powerful hurricanes and typhoons. Finally, positive feedback loops can cause further acceleration of global warming. For instance, melting of highly reflective polar ice can result in increased albedo (radiation absorption) by exposed ocean surface, further accelerating heating near the North Pole.

Potential risks from climate change stem from interactions with other critical sustainability issues, such as those listed by Rockström et al. (2009). As previously mentioned, climatic disruptions can lead to water stress if changes in seasonal precipitation levels lead to drought. Changes in ambient temperature can also lead to shifts in land use patterns. For example, it is entirely conceivable that entire agricultural regions in many countries will be rendered unsuitable to grow traditional, culturally accepted crops in future decades, as a result of changes in moisture and temperature levels. In such cases, adaptation measures such as shifting to alternative crops will be needed. Mass migration of human communities may become necessary as a result of such localized changes in weather, or from coastal inundation. Temperature shifts can also affect biodiversity of terrestrial and marine ecosystems. Effects on marine flora and fauna will also be compounded by ocean acidification resulting from higher levels of CO_2. Such damage can put even greater pressure on what are considered as *safe planetary boundaries* (Rockström et al., 2009). Table 1.1 lists the proposed planetary boundaries in comparison with prevailing values at the time; note that in the case of atmospheric CO_2 concentration, there has already been a significant increase in the years following the publication of their work.

There has been significant international concern about climate change since the late 20th century. In 2015, the 21st Conference of Parties (COP 21) of the United Nations Framework Convention on Climate Change (UNFCC) led to the Paris Agreement, which seeks to limit mean temperature rise to well below the high risk level of 2°C, through voluntary Nationally Determined Contributions (NDCs) committed by parties. This agreement represents significant progress from the earlier Copenhagen Agreement from COP 15 in 2009 and the Kyoto Protocol in 1997 (Held and Roger, 2018). Nevertheless, significant geopolitical barriers remain, as illustrated by the withdrawal of the United States from the Paris Agreement in 2017. Such decisions illustrate how narrow political interests might derail global efforts to curb climate change, despite a broad consensus on the need for action among most countries.

TABLE 1.1

Planetary Boundaries as Proposed by Rockström et al. (2009)

Earth System Process	Parameters	Proposed Boundary	2009 Level	Pre-Industrial Level
Climate change	CO_2 concentration (ppm)	350	387	280
	Change in radiative forcing (W/m^2)	1	1.5	0
Biodiversity loss	Extinction rate ($10^{-6}/y$)	10	>100	0.1–1
Nitrogen cycle	Net reactive N use by humans (Mt/y)	35	121	0
Phosphorus cycle	Net P flow into sea (Mt/y)	11	8.5–9.5	~1
Stratospheric ozone depletion	Concentration of ozone (Dobson unit)	276	283	290
Ocean acidification	Mean saturation state of aragonite in surface seawater	2.75	2.90	3.44
Freshwater use	Water consumption (km^3/y)	4,000	2,600	415
Land use change	Land cover converted to crop land (%)	15	11.7	Low
Atmospheric aerosol loading	Particulate concentration in the atmosphere	Not specified		
Chemical pollution	Quantity of various pollutants discharged into different compartments	Not specified		

In addition to the intent by nations, organizations, and individual persons to shift to climate-friendly behavior, technological enablers are essential. Some key technologies are discussed in the next section.

1.2 Technological Decarbonization Solutions

CO_2 accounts for a disproportionate share of radiative forcing when compared with other GHGs. Thus, many climate change mitigation measures focus specifically on reduction of CO_2 emissions, much of which is the result of the combustion of fossil fuels (coal, petroleum, and natural gas – NG). Measures to reduce carbon intensity of energy systems include the following:

- Shifting to lower-carbon fossil fuels (e.g., from coal to NG)
- Shifting to nuclear energy
- Shifting to renewable energy (e.g., solar, wind, hydroelectric, or biomass)
- Supply-side energy efficiency enhancement
- Demand-side energy conservation and efficiency enhancement
- *CO₂ capture, utilization, and storage* (CCUS)
- CO_2 removal (CDR) or *negative emissions technologies* (NETs)

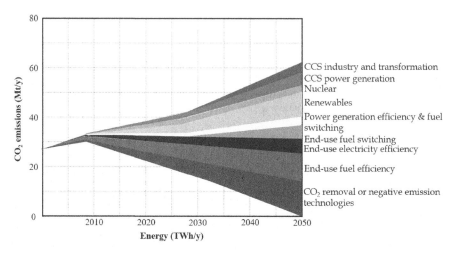

FIGURE 1.1
Technology wedges to reduce global GHG emissions.

The list above is by no means exhaustive, but it represents some of the most important technologies being considered. These techniques, and others not listed, may be deployed at scale as "technology wedges" (Figure 1.1; adapted from Pacala and Socolow, 2004) to contribute incrementally towards a global target for GHG emissions cuts; such an approach is more plausible than the assumption of a single technological solution. Since it has been estimated that deep cuts resulting in net zero emissions will be needed to achieve a climate trajectory of acceptable risk level (IPCC, 2018), it follows that CDR/NETs will be needed on a large scale to remove CO_2 from the atmosphere, and thus offset emissions generated elsewhere (Haszeldine et al., 2018).

Given the availability of many of these decarbonization technologies at present, or in the near future, decision makers will inevitably need to determine which measures to implement for specific locations and time frames. Such decisions will need to take into account various constraints such as economics, space, pollution levels, and social acceptability, among others. Decision support techniques will thus be essential to ensure that the best technologies are selected for implementation for any given context. The next section introduces one specific family of decision support tools.

1.3 Process Integration and Climate Change

Process integration emerged as a branch of chemical engineering in the 1970s to address efficiency concerns in energy-intensive process plants through optimal design of heat recovery system. Interest was driven largely by the global escalation of oil price and energy costs as a result of instability in the Middle East. At the time, alternative approaches based on *pinch analysis* and *mathematical programming* were developed as competing schools of thought. Pinch analysis is a two-stage, insight-based approach based on thermodynamic principles. Graphical displays are extensively used to determine system-level energy budgets (known in the process integration literature as *targets*) and to determine surplus and deficit zones in a given set of process units. Such information can then be used for subsequent design of *heat exchanger networks* (HENs) that achieve the predetermined targets. On the other hand, mathematical programming relies on representation of system components (e.g., heat exchangers, hot and cold streams, etc.) via algebraic equations and inequalities. Once a suitable performance measure, or *objective function*, is defined, a mathematical programming model can then be solved to optimality through the application of an appropriate algorithm.

Pinch analysis and mathematical programming approaches have their respective advantages and drawbacks. Pinch analysis methods require significant simplifying assumptions prior to use, and they may thus not be well-suited to dealing with detailed nuances of specific cases. On the other hand, they are inherently insight-based and are able to provide effective decision support by clearly indicating to the decision maker not only what the solution is but also *why* the solution is as it is. By comparison, mathematical programming models can be more readily customized to reflect various unique problem details. They are, however, also less transparent to the user. While a properly coded mathematical programming can, in principle, be relied upon to give a reliable solution, any errors in the model-building process are much harder to detect in the output solution. An important trend that has been observed in the evolution of process integration as an area of research is the increased appreciation of the complementary – rather than competing – roles that pinch analysis and mathematical programming can play as problem-solving strategies (Klemeš and Kravanja, 2013).

While process integration originally focused on *heat integration* problems for enhancing energy efficiency in process plants (Linnhoff et al., 1982), various extensions of this framework have been proposed. For example, the analogy between heat transfer and mass transfer led to the emergence of *mass integration* for the efficient use of industrial mass separating agents such as solvents (El-Halwagi and Manousiothakis, 1989). The diversity of process integration applications is documented in a comprehensive handbook with contributions from many of the field's most authoritative figures (Klemeš, 2013). A recent review focuses on pinch analysis developments (Klemeš et al., 2018).

Dhole and Linnhoff (1993) first proposed the potential of process integration as a means of reducing emissions of industrial sites. The emissions cuts they describe are the direct result of reduced fuel consumption due to increased heat recovery and improved efficiency. Significantly, their work predates the broader public understanding of climate change that occurred in the late 1990s and early 2000s. In the 21st century, Tan and Foo (2007) proposed process integration as a tool to facilitate decarbonization via the *carbon emissions pinch analysis* (CEPA) technique. In their seminal work (Tan and Foo, 2007), a graphical pinch analysis approach was proposed for the optimal matching of energy sources and demands, taking into account carbon intensity constraints. An equivalent linear programming (LP) model was also described. The method was applied to a simplified problem involving allocation of primary energy sources at the level of an entire country. Since its inception, the general CEPA methodology has been extended and adapted to different problem variants and geographical contexts, many of which are described in a review paper (Foo and Tan, 2016). The extension of CEPA to different measures of energy quality, such as land and water footprint, is described in a unified manner in a handbook chapter (Tan and Foo, 2013), while a reader-friendly introductory tutorial is given in an encyclopedia chapter (Tan and Foo, 2017). More recently, Tan and Foo (2018) proposed the term *carbon management networks* (CMNs) to collectively describe the application of process integration to the problem of climate change mitigation. They discussed briefly about the steady growth of specialist CEPA literature as a branch of process integration.[1] International interest in CEPA has been demonstrated by applications in different countries across six continents, including Ireland (Crilly and Zhelev, 2010), the United States of America (Walmsley et al., 2015), Brazil (de Lira Quaresma et al., 2018), Nigeria (Salman et al., 2018), the Baltic States of Estonia, Latvia, and Lithuania (Baležentis et al., 2019), as well as the countries represented in the chapters comprising the second half of this book. An international expert workshop was hosted by International Energy Agency (IEA) in year 2017 to examine the role of process integration and CEPA on GHG mitigation. An international expert workshop was hosted by International Energy Agency (IEA) in year 2017 to examine the role of process integration and CEPA on GHG mitigation. A recent article also examines the potential role of CEPA from a policy development standpoint (Andiappan et al., 2019). The rest of this book is intended to give a more detailed and accessible tutorial treatment of this area of study for the non-specialist reader. There are also five contributed chapters of case studies for China, New Zealand, India, the United Arab Emirates (UAE), Malaysia, and Taiwan, using CEPA and its extensions.

References

Andiappan, V., Foo, D. C. Y., Tan, R. R. 2019. Process-to-Policy (P2Pol): Using carbon emission pinch analysis (CEPA) tools for policy-making in the energy sector. *Clean Technologies and Environmental Policy*, 21, 1383–1388.

---

[1] As of end of 2019, the original CEPA paper (Tan and Foo, 2017) has been cited more than 200 times in the Scopus database.

Baležentis, T., Štreimikienė, D., Melnikienė, R., Zeng, S. 2019. Prospects of green growth in the electricity sector in Baltic States: Pinch analysis based on ecological footprint. *Resources, Conservation and Recycling*, 142, 37–48.

Crilly, D., Zhelev, T. 2010. Further emissions and energy targeting: An application of CO_2 emissions pinch analysis to the Irish electricity generation sector. *Clean Technologies and Environmental Policy*, 12, 177–189.

Dhole, V. R., Linnhoff, B. 1993. Total site targets for fuel, co-generation, emissions and cooling. *Computers and Chemical Engineering*, 17, s101–s109.

El-Halwagi, M. M., Manousiouthakis, V. (1989). Synthesis of mass exchange networks. AIChE Journal. 35(8): 1233–1244.

Foo, D. C. Y., Tan, R. R. 2016. A review on process integration techniques for carbon emissions and environmental footprint problems. *Process Safety and Environmental Protection*, 103, 291–307.

Haszeldine, R. S., Flude, S., Johnson, G., Scott, V. 2018. Negative emissions technologies and carbon capture and storage to achieve the Paris Agreement commitments. *Philosophical Transactions of the Royal Society A: Mathematical, Physical and Engineering Sciences*. doi: 0.1098/rsta.2016.0447.

Held, D., Roger, C. 2018. Three models of global climate governance: From Kyoto to Paris and beyond. *Global Policy*, 9, 527–537.

IEA, 2017. https://iea-industry.org/activities/event/iea-expert-workshop-the-role-of-process-integration-for-greenhouse-gas-mitigation-in-industry/

IPCC, 2018. Summary for policymakers. In Masson-Delmotte, V., Zhai, P., Pörtner, H.O., Roberts, D., Skea, J., Shukla, P. R., Pirani, A., Moufouma-Okia, W., Péan, C., Pidcock, R., Connors, S., Matthews, J. B. R., Chen, Y., Zhou, X., Gomis, M. I., Lonnoy, E., Maycock, T., Tignor, M., Waterfield, T. (Eds.), *Global Warming of 1.5°C*. An IPCC Special Report on the impacts of global warming of 1.5°C above pre-industrial levels and related global greenhouse gas emission pathways, in the context of strengthening the global response to the threat of climate change, sustainable development, and efforts to eradicate poverty. World Meteorological Organization, Geneva, Switzerland.

Klemeš, J.J. (Ed.) 2013. *Process Integration Handbook*. Woodhead Publishing/Elsevier, Cambridge, UK.

Klemeš, J.J., Kravanja, Z. 2013. Forty years of heat integration: pinch analysis (PA) and mathematical programming (MP). *Current Opinion in Chemical Engineering*, 2, 461–474.

Klemeš, J. J., Varbanov, P. S., Walmsley, T. G., Jia, X. 2018. New directions in the implementation of Pinch Methodology (PM). *Renewable and Sustainable Energy Reviews*, 98, 439–468.

Linnhoff, B., Townsend, D. W., Boland, D., Hewitt, G. F., Thomas, B. E. A., Guy, A. R., Marshall, R. H. 1982. *A User Guide on Process Integration for the Efficient Use of Energy*. Institute of Chemical Engineers, Rugby, UK.

de Lira Quaresma, A. C., Francisco, F. S., Pessoa, F. L. P., Queiroz, E. M. 2018. Carbon emission reduction in the Brazilian electricity sector using Carbon Sources Diagram. *Energy*, 159, 134–150.

Pacala, S., Socolow, R. 2004. Stabilization wedges: Solving the climate problem for the next 50 years with current technologies. *Science*, 305, 968–972.

Rockström, J., Steffen, W., Noone, K., Persson, A., Chapin, F. S., Lambin, E. F., Lenton, T. M., Scheffer, M., Folke, C., Schellnhuber, H. J., Niykvist, B., De Wit, C. A., Hughes, T., Van der Leeuw, S., Rodhe, H., Sorlin, S., Snyder, P. K., Constanza,

R., Svedin, U., Falkenmark, M., Karlberg, L., Corell, R. W., Fabry, V. J., Hansen, J., Walker, B., Liverman, D., Richardson, K., Crutzen, P., Foley, J. A. 2009, A safe operating space for humanity. *Nature*, 461, 472–475.

Tan, R. R., Foo, D. C. Y. 2007, Pinch Analysis approach to carbon-constrained energy sector planning. *Energy*, 32, 1422–1429.

Tan, R. R., Foo, D. C. Y. 2013, Pinch Analysis for sustainable energy planning using diverse quality measures. In Klemeš, J. J. (Ed.), *Handbook of Process Integration (PI): Minimisation of Energy and Water Use, Waste and Emissions*. Elsevier/Woodhead Publishing, Cambridge, UK, pp. 505–523.

Tan, R. R., Foo, D. C. Y. 2017, Carbon Emissions Pinch Analysis for sustainable energy planning. In Abraham, M. (Ed.), *Encyclopedia of Sustainable Technologies*. Elsevier, Amsterdam, pp. 231–237.

Tan, R.R., Foo, D. C. Y. (2018). Process integration and climate change: From carbon emissions pinch analysis to carbon management networks. Chemical Engineering Transactions, 70: 1–6.

Salman, B., Nomanbhay, S., Foo, D. C. Y. 2018. Carbon emissions pinch analysis (CEPA) for energy sector planning in Nigeria. *Clean Technologies and Environmental Policy*, 21, 93–108.

Walmsley, M. R. W., Walmsley, T. G., Atkins, M. J. 2015. Achieving 33% renewable electricity generation by 2020 in California. *Energy*, 92, 260–269.

2

Graphical Targeting Techniques for Carbon Emission Pinch Analysis (CEPA)

This chapter outlines an important graphical-based targeting tool based on *carbon emission pinch analysis* (CEPA). This method is similar to other graphical techniques in traditional pinch analysis applications such as energy (Smith, 2016) and mass integration (El-Halwagi, 2017), where sources and demands are plotted as *composite curves*, whose relative geometric orientations reflect the quantity and quality constraints of the system. The main philosophy of CEPA tools is to set performance targets for the system. For this case, the performance targets correspond to minimum amount of renewable sources, and the excess amount of conventional fossil fuels. Note that renewable sources are usually scarce and relatively more expensive than fossil fuels, hence their uses are minimized. The graphical approach also provides an intuitively appealing visual representation, which is useful for providing essential insights to support decision-making and communication of results.

2.1 Graphical Tool – Energy Planning Pinch Diagram (EPPD)

An important graphical technique known as *energy planning pinch diagram* (EPPD) is outlined here. The EPPD relies on the fact that carbon footprint follows a *linear mixing rule* – i.e., the carbon footprint of a mix of multiple independent energy sources is simply the sum of the footprints of these sources taken individually. From this principle, it also follows that the carbon intensity of the energy mix is the weighted average of the intensity of the components of the energy mix. In addition, it is clear that lower numerical values of carbon footprint are indicative of better quality. These properties allow the application of pinch analysis to energy planning problems.

The procedure for plotting the EPPD is outlined as follows:

   i. All energy demands and sources are arranged respectively in ascending order of their carbon intensity.
   ii. The energy demands are plotted in a CO_2 load vs. energy diagram to form the *demand composite curve*, in order of ascending carbon intensity.

iii. The energy sources are plotted in a CO_2 load vs. energy diagram to form the *source composite curve*, in order of ascending carbon intensity.

iv. The demand and source composite curves are superimposed in the same CO_2 load vs. energy diagram to form the EPPD.

v. In cases where the source composite curve is entirely below and to the right of the demand composite curve, and provided the horizontal span of the former is sufficient to cover the total energy demand, then the initial solution is immediately feasible.

vi. For cases where the source composite curve crosses to the left or above the demand composite curve, the EPPD is infeasible. To obtain a feasible EPPD, additional carbon-neutral or low-carbon energy resource is to be used. The amount required is determined during the application of the procedure. For cases where carbon-neutral resource (i.e., with carbon intensity, $C_{RE} = 0$) is available, the source composite curve is to be shifted horizontally to the right. When low-carbon energy resource is used, on the other hand, the source composite curve is to be shifted along a locus, whose slope corresponds to its carbon intensity. For both cases, the source composite curve is shifted until it lies entirely to the right and below the demand composite curve, and touches the latter at the *pinch*.

Two examples will be used to elucidate the graphical targeting tool.

Example 2.1 Graphical Targeting Technique for Illustrative Example (Tan and Foo, 2017)

An illustrative example from Tan and Foo (2017) is used to illustrate the plotting of EPPD. Data for this example are shown in Table 2.1. The table shows two energy demands, each with its energy requirement and CO_2 emission limit. The latter may be divided by the energy requirement to determine the corresponding carbon intensity of the demand. On the other hand, available for use are two energy sources, with their

TABLE 2.1

Data for Example 2.1: (a) Energy Demands and (b) Energy Sources (Tan and Foo, 2017)

(a) Demand j	Energy, $F_{D,j}$ (TWh/y)	Carbon Intensity, $C_{D,j}$ (Mt/TWh)	CO_2 Emission, $E_{D,j}$ (Mt/y)
D1	200	0.25	50
D2	100	1.00	100
(b) Source i	Energy, $F_{S,i}$ (TWh/y)	Carbon Intensity, $C_{S,i}$ (Mt/TWh)	CO_2 Emission, $E_{S,i}$ (Mt/y)
S1	200	0.5	100
S2	150	1.5	225

availability and carbon intensity. The product of energy and carbon intensity determines the CO_2 emission of the respective source. When energy sources are insufficient for use, carbon-neutral energy resource (i.e., $C_{RE} = 0$) is to be used.

Note that the energy demands and sources given in Table 2.1 have already been arranged in order of ascending carbon intensity. Hence, we proceed directly to steps ii and iii of the EPPD procedure outlined in the previous section, to generate the demand and source composite curves, respectively (see Figure 2.1). The demand composite curve (Figure 2.1a) shows that the two regions require a total energy requirement of 300 TWh/y, and has a total emission limit of 150 Mt/y CO_2. The source composite curve, on the other hand, indicates that the two available sources have a

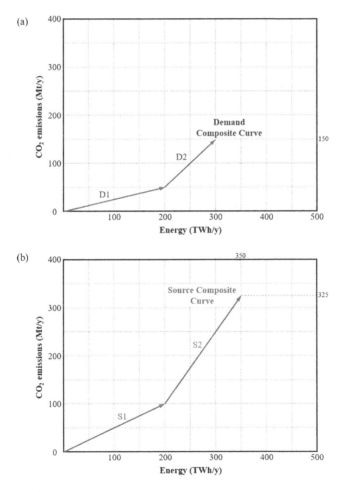

FIGURE 2.1
The plotting of (a) demand composite curve and (b) source composite curve (Tan and Foo, 2017).

total energy of 350 TWh/y, with a total emission of 325 Mt/y. By inspection, there is a surplus of both energy (50 = 350–300 TWh/y) and CO_2 emissions (175 = 325–150 Mt/y).

Next, the EPPD is constructed by superimposing the demand and source composite curves on the same diagram (step iv), as shown in Figure 2.2. However, the figure shows that the source composite curve is to the left and above the demand composite curve, which means that EPPD is infeasible. In particular, the two energy demands have a total CO_2 emission limit of 150 Mt/y; however, emissions will amount to 250 Mt/y of CO_2 if energy is drawn only from these two sources. Thus, supplementary zero-carbon or carbon-neutral energy is needed to ensure that emissions limits are met.

Step v of the EPPD procedure is then followed. The source composite curve is moved horizontally to the right, until it stays completely below and to the right of the demand composite curve, and just touches the latter at the pinch. This results in the feasible EPPD (Figure 2.3). For the latter, the horizontal gap on the left indicates the minimum amount of carbon-neutral energy resource (F_{RE}) that needs to be added to the system, i.e., 100 TWh/y. This carbon-neutral energy resource can consist of technologies with carbon footprints that are negligible in comparison to the fossil energy sources, such as renewables or nuclear power. On the other hand, the opening on the right of the EPPD indicates the excess energy (F_{EX}) that cannot be utilized, i.e., 150 TWh/y (= 450–300 TWh/y), if CO_2 emission limit is to be met. Note that the entire segment of the source composite curve corresponding to S2 lies beyond the horizontal span of the demand composite curve, indicating that none of it can be used in the system. It can also be seen that D1 is below the pinch point, while D2 is above it. The significance of these positions relative to the pinch point will be discussed in a later section in this chapter.

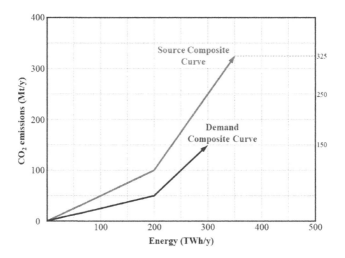

FIGURE 2.2
An infeasible EPPD for Example 2.1 (Tan and Foo, 2017).

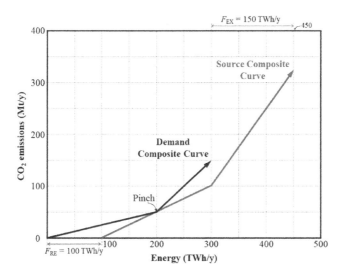

FIGURE 2.3
A feasible EPPD for Example 2.1 (Tan and Foo, 2017).

Example 2.2 Graphical Targeting Technique for Classical Example (Tan and Foo, 2007)

In this classical example from Tan and Foo (2007), three demands are in need of energy resources. Three fossil fuels may be utilized for the demand, which is supplemented by carbon-neutral (i.e., $C_{RE} = 0$) and low-carbon energy resources (i.e., $C_{RE} = 16.5$ t/TJ). Data for the example are summarized in Table 2.2.

Following steps i–iv of the EPPD procedure, the infeasible EPPD is generated in Figure 2.4, with the source composite curve lying to the left and above the demand composite curve.

TABLE 2.2

Data for Example 2.2: (a) Energy Demands and (b) Energy Sources (Tan and Foo, 2007)

(a) Demand j	Energy, $F_{D,j}$ (TJ)	Carbon Intensity, $C_{D,j}$ (t/TJ)	CO$_2$ Emission, $E_{D,j}$ (t)
D1	1,000,000	20	20,000,000
D2	400,000	50	20,000,000
D3	600,000	100	60,000,000

(b) Source i	Energy, $F_{S,i}$ (TJ)	Carbon Intensity, $C_{S,i}$ (t/TJ)	CO$_2$ Emission, $E_{S,i}$ (t)
S1	200,000	55	11,000,000
S2	800,000	75	60,000,000
S3	600,000	105	63,000,000

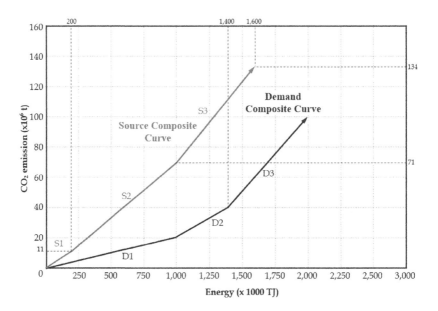

FIGURE 2.4
An infeasible EPPD for Example 2.2 (Tan and Foo, 2007).

As in the previous example, when carbon-neutral resource is utilized, the source composite curve is shifted to the right horizontally, until it is entirely to the right and below the demand composite curve (step v). This results with the feasible EPPD, as in Figure 2.5. The opening on the left of the EPPD indicates that a total of 813,000 TJ of carbon-neutral energy (F_{RE}) is needed, in order to keep the emissions in all regions below their CO_2 emission limits. On the other hand, the opening on the right of the EPPD shows the excess energy (F_{EX}) of 413,000 TJ should not be utilized, if the CO_2 emission limit is to be met. The unused energy source is the one with the highest carbon intensity among available options – i.e., coal.

On the other hand, when a low-carbon resource is utilized, a locus is added in the EPPD, whose slope corresponds to its carbon intensity, (C_{RE} = 16.5 t/TJ). In typical CEPA problems, this carbon intensity will be lower than that of the other energy sources, so the locus assumes the form of a shallow ramp along which the source composite curve can be shifted diagonally. The source composite curve is thus shifted to the right along this locus, until it is entirely on the right and below the demand composite curve (step v). The resulting feasible EPPD is shown in Figure 2.6. For this case, the low-carbon energy (F_{RE}) is determined as 1,040,000 TJ, while 640,000 TJ (= 2,640,000–2,000,000 TJ) of excess energy sources are left unutilized (F_{EX}). Both of these quantities are notably larger than what would be needed if carbon-neutral energy is used, as previously shown.

It can be seen in the case of both carbon-neutral and low-carbon energy resource that D1 and D2 are both below the pinch point, while D3 is above the pinch point. The significance of their positions relative to the pinch are discussed in the next section.

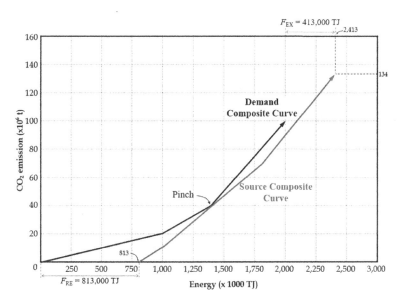

FIGURE 2.5
A feasible EPPD for Example 2.2 with carbon-neutral resource (Tan and Foo, 2007).

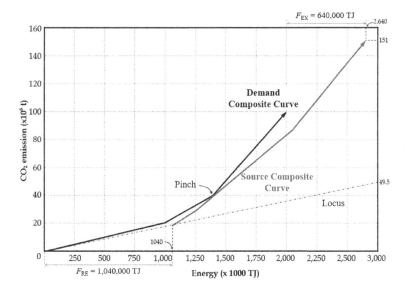

FIGURE 2.6
A feasible EPPD for Example 2.2 with low-carbon resource (Lee et al., 2009).

2.2 Significance of the Pinch Point

The significance of the pinch point in CEPA is analogous to the role of pinch in other process integration applications (e.g., energy (Smith, 2016) and mass integration (El-Halwagi, 2017)). The pinch point divides the system into below-pinch and above-pinch regions (not to be confused with geographic regions that may be represented by demand composite curve segments). The pinch point decomposes the overall problem into two sub-problems that may be analyzed separately. In fact, in the optimal solution, energy allocation must respect the so-called "golden rule of pinch analysis," which forbids any transfer of energy across the pinch point from one region to the other.

Below the pinch point, the CO_2 limits of the demands are met exactly, without any degrees of freedom. There is only one way in which energy sources below the pinch (including the carbon-neutral and low-carbon resources) can be allocated in order to ensure that each demand's emissions limit is met. The lack of excess degrees of freedom can be visualized by the two composite curves emanating from the same point (i.e., the origin of EPPD), diverging, and then converging again at the pinch point. This geometry means that the total energy and total CO_2 emissions of the sources exactly match those of the demands.

Above the pinch point, it can be seen that the two composite curves generally diverge without meeting again. It is necessary to ensure that the source composite curve extends far enough to the right to cover the energy requirements represented by the demand composite curve. Any portion that extends beyond the demand composite curve represents a surplus which is not utilized. However, an important feature of the above-pinch region is the availability of excess degrees of freedom, signified by the vertical gap between the composite curves, which can allow for alternative allocation schemes that still meet the emissions limits of the different (Tan and Foo, 2007).

Depending on the shapes of the composite curves, special conditions may also arise. For example, it is possible to have two or more pinch points in a system. In such a scenario, the conditions discussed above can be generalized. Cross-pinch transfer is still forbidden, and all regions separated by pinch points can be analyzed and solved independently of each other. In addition, all regions bounded between pinch points have zero degrees of freedom, just like the below-pinch region in the general case. There is exactly one way to allocate energy sources to the demands in any region between two pinch points.

If the horizontal span of the source composite is particularly short, it may be necessary to shift it farther to the right, past the point where a pinch is formed, in order to ensure that the energy balance condition is met. Such a case is known as a *threshold problem*, and the entire system behaves as a single below-pinch region.

2.3 Extension for Other Environmental Footprints

The principles of CEPA can be extended to other sustainability measures that conform to the linear mixing assumption defined earlier in this chapter. Different applications and case studies are discussed in a book chapter (Tan and Foo, 2013).

Water footprint is the cumulative water requirement associated with a product over its entire life cycle. Energy systems have water footprints that may be relevant to sustainable energy planning, because disruptions in water supply (e.g., drought) can have an impact on energy supply. This link has led to the concept of the water-energy nexus. For example, water is needed in large quantities for steam generation and for cooling in thermal power plants, including those running on fossil fuels, nuclear fuel, and biomass. Even renewable energy systems can have significant water footprints as well. Examples include evaporative losses in hydroelectric power, and water demand for energy crop cultivation in bioenergy systems. Water footprint is also indicative of potential vulnerability of energy systems to water shortage or drought.

Land footprint is a measure of the land area used up by an energy system. This may be critical in the case of densely populated regions or countries where land is also essential for agricultural, industrial, and residential use. Examples of land footprint occurring from energy systems are the land requirements of solar energy farms and biomass plantations, as well as land area flooded upstream of large hydroelectric power plants.

An example is considered here that is relevant to energy planning.

Example 2.3 Graphical Targeting Technique for Water Footprint Example

This example considers the hypothetical case of an island with two regions, D1 and D2. Each region has known energy demand and water footprint limit, the latter being based on the average amount of rainfall within the territory. The island is projected to have 30 TWh/y of hydroelectricity potential (S2). Thus, additional power generation capacity from coal-fired plants (S1) is required. The latter capacity needs to be minimized in order to keep greenhouse gas (GHG) emissions minimal. The data for demands and sources are given in Table 2.3.

TABLE 2.3

Data for Example 2.3: (a) Energy Demands and (b) Energy Sources

(a) Demand $_j$	Energy, $F_{D,j}$ (TWh/y)	Water Footprint Intensity, $W_{D,j}$ (Mt/TWh)	Water Footprint Limit, $E_{D,j}$ (Mt/y)
D1	20	100	2,000
D2	15	367	5,505

(b) Source i	Energy, $F_{S,i}$ (TWh/y)	Water Footprint Intensity, $W_{S,i}$ (Mt/TWh)	Water Footprint Limit, $E_{S,i}$ (Mt/y)
S1	To be determined	4	To be determined
S2	30	245	7,350

Water footprint also conforms to the linear mixing rule mentioned earlier in this chapter. Thus, the EPPD procedure can be applied using water footprint intensity as the quality indicator. Applying the same procedure as in Example 2.2, the composite curves can be generated as shown in Figure 2.7. Initially it is assumed that zero coal-based capacity is planned. The resulting EPPD shows infeasible orientation, where the source composite curve lies on the left of the demand composite curve as seen in Figure 2.7. This result means that coal-fired power plants will have to be built to meet the energy deficit without exceeding the water resources in the regions. Thus, the source composite curve is shifted to the right along a locus (with slope corresponds to a water intensity of 4 Mt/TWh), until a feasible and optimal orientation is found, as shown in Figure 2.8. It can be seen that the minimum required coal-fired power generation is 12 TWh/y.

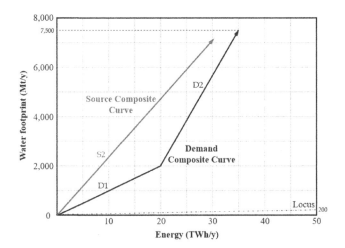

FIGURE 2.7
An infeasible EPPD for Example 2.3.

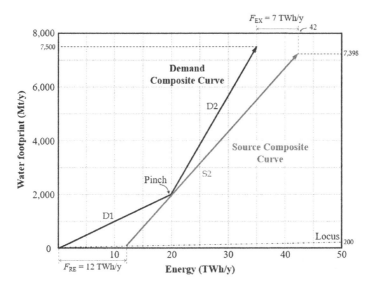

FIGURE 2.8
A feasible EPPD for Example 2.3 with low-water footprint resource.

It can also be seen that this capacity from coal-fired power plants falls below the pinch point. Following the golden rule of pinch analysis, this result means that all of this capacity will be installed in the first region (D1), along with 8 TWh/y (= 20–12 TWh/y) of hydroelectric power (S2). The use of the coal-fired power plant thus puts less stress on the relatively limited water resources in this region. By comparison, the entire 15 TWh/y of demand in the second region (D2) can be met with just hydroelectric power (S2). It can also be seen that this optimal allocation does not use up the entire hydroelectric potential of the island, but leaves a surplus that can be tapped for future expansion. The latter is determined as 7 TWh/y (F_{EX}), as shown in Figure 2.8.

Further Reading

Alternative graphical pinch analysis methods such as the source composite curve developed by Bandyopadhyay (2006) can also be readily adapted to CEPA and related problems. The equivalence between EPPD and the alternative method is discussed by Bandyopadhyay (2015). For a comprehensive discussion of different footprint-based sustainability metrics, the reader may refer to the review paper by Čuček et al. (2012). Readers may also refer to the works of Jia et al. (2016) and Patole et al. (2017) for multidimensional CEPA.

References

Bandyopadhyay, S. 2006. Source composite curve for waste reduction. *Chemical Engineering Journal*, 25(2), 99–110.

Bandyopadhyay, S. 2015. Mathematical foundation of pinch analysis. *Chemical Engineering Transactions*, 45, 1753–1758.

Čuček, L., Klemeš, J. J., Kravanja, Z. 2012. A review of footprint analysis tools for monitoring impacts on sustainability. *Journal of Cleaner Production*, 34(1), 9–20.

El-Halwagi, M. M. 2017. *Sustainable Design through Process Integration*, 2nd Ed. Elsevier, Amsterdam.

Jia, X., Li, Z., Wang, F., Foo, D. C. Y., Tan, R. R. 2016. Multi-dimensional pinch analysis for power generation sector in China. *Journal of Cleaner Production*, 112, 2756–2771.

Lee, S. C., Ng, D. K. S., Foo, D. C. Y., Tan, R. R. 2009. Extended pinch targeting techniques for carbon-constrained energy sector planning. *Applied Energy*, 86(1), 60–67.

Patole, M., Bandyopadhyay, S., Foo, D. C. Y., Tan, R. R. 2017. Energy sector planning using multiple-index pinch analysis. *Clean Technologies and Environmental Policy*, 19, 1967–1975.

Smith, R. 2016. *Chemical Process: Design and Integration*, 2nd Ed. John Wiley, West Sussex.

Tan, R. R., Foo, D. C. Y. 2007. Pinch analysis approach to carbon-constrained energy sector planning. *Energy*, 32(8), 1422–1429.

Tan, R. R., Foo, D. C. Y. 2013. Pinch analysis for sustainable energy planning using diverse quality measures. In Klemeš, J. (Ed.), *Handbook of Process Integration (PI): Minimisation of Energy and Water Use, Waste and Emissions*. Elsevier/Woodhead Publishing, Cambridge, UK, pp. 505–523.

Tan, R. R., Foo, D. C. Y. 2017. Energy sector planning with footprint constraints. In Abraham, M. A. (Ed.), *Encyclopedia of Sustainable Technologies*. Elsevier, Amsterdam, pp. 231–237.

3

Algebraic and Automated Targeting Techniques for Carbon Emission Pinch Analysis (CEPA)

The graphical targeting technique in Chapter 2 serves as a good visualization tool for users, as it provides intuitive appealing visual representation, which is useful in providing insights as well as the communication of results. However, all graphical techniques suffer from limitations such as inaccuracy and being cumbersome. This chapter outlines several other pinch-based targeting techniques, which may be used to supplement the graphical technique in Chapter 2. Algebraic targeting tool is first presented; which is shown to have higher accuracy as compared to the graphical tool. The algebraic targeting tool may be extended into *automated targeting model* (ATM), an optimization framework that can be implemented through spreadsheet and commercial optimization software. The ATM allows minimum energy and cost solutions to be determined.

3.1 Algebraic Targeting Technique

Table 3.1 shows the basic structure for algebraic targeting technique. Individual steps to carry out the analysis are given as follows (Foo et al., 2008):

i. In column 1, carbon intensity (C_k) for all energy demands and sources are arranged in ascending order. An arbitrary large value is added at the last level n to facilitate calculation. The difference between adjacent levels of intensity (ΔC_k) is then calculated in column 2.

ii. In columns 3 and 4, energy demands and sources are summed and located at their respective level.

iii. In column 5, the *net energy* for each level k ($F_{\text{Net}, k}$) is determined from the difference between the total energy sources with that of the energy demands, given as in Equation 3.1.

$$F_{\text{Net}, k} = \sum_i F_{\text{S}, i} - \sum_j F_{\text{D}, j} \quad \forall k \qquad (3.1)$$

TABLE 3.1

Basic Structure for Algebraic Targeting Technique

C_k	ΔC_k	$\Sigma_j F_{D,j}$	$\Sigma_i F_{S,i}$	$F_{Net,k}$	Cum. $F_{Net,k}$	ΔE_k	Cum. ΔE_k	$F_{RE,k}$
					F_{RE}			
C_1		$\Sigma_j F_{D,j}$	$\Sigma_i F_{S,i}$	$F_{Net,1}$	⇓		Cum. $\Delta E_1 = 0$	$F_{RE,1} = 0$
	ΔC_1				Cum. $F_{Net,1}$	ΔE_1	⇓	
C_2		$\Sigma_j F_{D,j}$	$\Sigma_i F_{S,i}$	$F_{Net,2}$	⇓		Cum. ΔE_2	$F_{RE,2}$
⋮	ΔC_2	⋮	⋮	⋮	Cum. $F_{Net,2}$	ΔE_2	⋮	⋮
⋮	⋮	⋮	⋮	⋮	⋮	⋮	⋮	⋮
C_{n-1}	⋮	$\Sigma_j F_{D,j}$	$\Sigma_i F_{S,i}$	$F_{Net,2}$	⋮	⋮	Cum. ΔE_{n-1}	$F_{RE,n-1}$
	ΔC_{n-1}				Cum. $F_{Net,n-1}$	ΔE_{n-1}	⇓	
C_n					(= F_{EX})		Cum. ΔE_n	$F_{RE,n}$

iv. In column 6, the net energy is added to that at the next level to obtain the *cumulative energy* at interval k (Cum. $F_{Net,k}$), which then forms the *energy cascade*. The last entry of this column indicates the *excess energy* (F_{EX}) of the system.

v. In column 7, the *emission load* for each interval k (ΔE_k) is calculated; the latter is given as the product of cumulative energy with two adjacent intensity levels, following Equation 3.2.

$$\Delta E_k = \text{Cum. } F_{Net,k} \left(C_{k+1} - C_k \right) \quad \forall k \tag{3.2}$$

vi. In column 8, the *cumulative emission load* of level k (Cum. ΔE_k) is calculated following Equation 3.3. This forms the *emission load cascade*. For a feasible load cascade, there should be no negative Cum. ΔE_k value in this column. In other words, the Cum. ΔE_k value can only take positive value, or zero; the latter indicates a pinch point.[1]

$$\text{Cum. } \Delta E_k = \begin{cases} 0 & k = 1 \\ \Delta E_{k-1} + \text{Cum. } \Delta E_{k-1} & k \geq 2 \end{cases} \tag{3.3}$$

When negative values are observed for Cum. ΔE_k, go to steps vii and viii, else to step ix.

vii. When there are negative Cum. ΔE_k values being observed in column 8, the emission load cascade is considered infeasible. Hence, carbon-neutral or low-carbon (renewable) energy is to be added at the first entry in column 6. In order to determine the amount of renewable energy needed for each level k ($F_{RE,k}$), Equation 3.4 is to be used. In principle, the renewable energy needed for each level

---

[1] See Chapter 2 for detailed discussion on the significance of pinch point.

k is determined by dividing the Cum. ΔE_k values (in column 8) by the difference between intensity level k with that of the renewable energy; the calculated values are summarized in column 9.

$$F_{RE,\,k} = \begin{cases} 0 & k = 1 \\ \dfrac{\text{Cum. } \Delta E_k}{C_k - C_{RE}} & k \geq 2 \end{cases} \tag{3.4}$$

viii. The absolute value of the minimum (i.e., largest deficit) value among those in column 9 is taken as the renewable energy target for the system (F_{RE}), following Equation 3.5. This amount of renewable energy is to be used for the first entry in the energy cascade in column 6. Steps iv–vi are repeated to construct the *feasible load cascade*.

$$F_{RE} = \left| \min_k F_{RE,k} \right| \tag{3.5}$$

ix. For cases with negative value observed for F_{EX}, the energy cascade is *infeasible*. Additional energy (in the form of renewables) needs to be added to the system to restore the feasibility. The amount of renewable energy (F_{RE}) to be added is given as the absolute value of F_{EX}, following Equation 3.6. The energy cascade is reconstructed using this renewable energy target (F_{RE}) that is located as the first entry.

$$F_{RE} = |F_{EX}| \tag{3.6}$$

The overall procedure is summarized in a flowchart in Figure 3.1.

Example 3.1 Algebraic Targeting Technique for Illustrative Example (Tan and Foo, 2017)

Example 2.1 is revisited to demonstrate the algebraic targeting technique. Data for the example are found in Table 2.1. Following steps i–ii, the energy demands and sources are located according to their respective carbon intensity levels; the latter has been arranged in ascending order, with an arbitrary large value ($C_n = 10$) added at the last level. The net energy is then calculated for each level k ($F_{Net,\,k}$), following step iii. In step iv, the energy cascade is constructed from the cumulative energy values across the intervals. The last entry of the energy cascade indicates that the system has an excess energy of 50 TWh/y.

We next proceed to calculate the emission load for each interval k (step v) and then to construct the emission load cascade (step vi). As shown in Table 3.2(a), the first three entries in the emission load cascade are observed to have negative values, i.e., an infeasible load cascade. We then move to steps vii and viii to identify the renewable energy target (F_{RE}); the latter corresponds to the largest deficit value identified at the intensity level of 0.5 Mt/TWh, i.e., 100 TWh/y. Steps iv–vi are then repeated

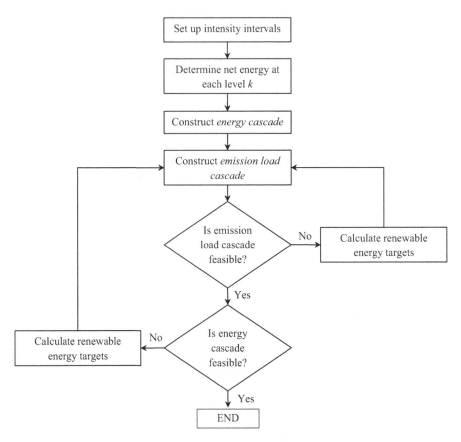

FIGURE 3.1
Flowchart for algebraic targeting procedure.

to construct the feasible load cascade given as in Table 3.2(b). The latter also shows that the excess energy (F_{EX}) is identified from the last entry in column 5, i.e., 150 TWh/y. Both renewable and excess energy targets are identical to those found using the graphical technique (i.e., *energy planning pinch diagram* – EPPD) in Chapter 2 (see Figure 2.3). Table 3.2(b) also shows that the pinch point (C_{PN}) is located from the intensity level of 0.5 Mt/TWh, where zero cumulative emission load is found.

Example 3.2 Algebraic Targeting with Spreadsheet for Classical Example (Tan and Foo, 2007) 🖥

The algebraic targeting technique is conveniently implemented in a spreadsheet tool. Figure 3.2 shows the snapshot for such implementation on Microsoft Excel, for a classical example from Tan and Foo (2007). Data for the example are found in Table 2.2. For this case, it is assumed that carbon-neutral energy resource is used, i.e., with no carbon intensity ($C_{RE} = 0$). As shown in Figure 3.2, steps i–iii of the targeting technique are carried out in columns 1–5 of the spreadsheet. Columns 6–9 of the spreadsheet show the infeasible cascades (steps iv–vi) as negative values are observed in the emission load cascade. Hence, steps vii–viii are followed to determine the minimum renewable target of the system, as implemented in columns 10–12 of the spreadsheet in Figure 3.2. The minimum renewable and minimum excess energy targets are determined as 813,333 and 413,333 TJ, given, respectively, in the first and last entries in column 10 (Figure 3.2).

TABLE 3.2

Algebraic Targeting for Example 3.1: (a) Infeasible Cascade and (b) Feasible Cascade

C_k (Mt/ TWh)	ΔC_k (Mt/ TWh)	$\Sigma_j F_{D,j}$ (TWh/y)	$\Sigma_i F_{S,i}$ (TWh/y)	$F_{Net,k}$ (TWh/y)	Cum. $F_{Net,k}$ (TWh/y)	ΔE_k (Mt/y)	Cum. ΔE_k (Mt/y)	$F_{RE,k}$ (TWh/y)
(a)								
					0			
0								
	0.25				0			
0.25		200		−200				
	0.25				−200	−50		
0.50			200	200			−50	−100
	0.50				0	0		
1.00		100		−100			−50	−50
	0.50				−100	−50		
1.50			150	150			−100	−66.7
	8.50				50	425		
10.00					(F_{EX})		325	32.5
C_k (Mt/ TWh)	ΔC_k (Mt/ TWh)	$\Sigma_j F_{D,j}$ (TWh/y)	$\Sigma_i F_{S,i}$ (TWh/y)	$F_{Net,k}$ (TWh/y)	Cum. $F_{Net,k}$ (TWh/y)	ΔE_k (Mt/y)	Cum. ΔE_k (Mt/y)	
(b)								
					100			
0					(F_{RE})			
	0.25				100	25		
0.25		200		−200			25	
	0.25				−100	−25		
0.50			200	200			0	
(C_{PN})	0.50				100	50		
1.00		100		−100			50	
	0.50				0	0		
1.50			150	150			50	
	8.50				150	1,275		
10.00					(F_{EX})		1,325	

C_t (t/TJ)	ΔC (t/TJ)	$\Sigma_j F_{Dj}$ (TJ)	$\Sigma_i F_{Si}$ (TJ)	$F_{Net, t}$ (TJ)	Infeasible cascade Cum. $F_{Net, t}$ (TJ)	ΔE_t (t)	Cum. ΔE_t (t)	$F_{CS, t}$ (TJ)	Feasible cascade Cum. $F_{Net, t}$ (TJ)	ΔE_t (t)	Cum. ΔE_t (t)
0				0	0				813,333	Minimum renewable target	
	20				0	0			813,333	16,266,667	
20		1,000,000		-1,000,000			0	0			16,266,667
	30				-1,000,000	-30,000,000			-186,667	-5,600,000	
50		400,000		-400,000			-30,000,000	-600,000			10,666,667
	5				-1,400,000	-7,000,000			-586,667	-2,933,333	
55			200,000	200,000			-37,000,000	-672,727			7,733,333
	20				-1,200,000	-24,000,000			-386,667	-7,733,333	
75			800,000	800,000			-61,000,000	-813,333			0 (Pinch)
	25				-400,000	-10,000,000			413,333	10,333,333	
100		600,000		-600,000			-71,000,000	-710,000			10,333,333
	5				-1,000,000	-5,000,000			-186,667	-933,333	
105			600,000	600,000			-76,000,000	-723,810			9,400,000
	4895				-400,000	-1,958,000,000			413,333	2,823,266,667	
5000							-2,034,000,000	-406,800		Minimum excess energy	2,032,666,667

FIGURE 3.2
Algebraic targeting technique with MS Excel (with carbon-neutral renewables).

One may also compare the targets found using the graphical technique in Chapter 2. The EPPD in Figure 2.5 shows less accurate targets due to its graphical nature. This limitation is now overcome by the spreadsheet tool.

We next demonstrate the case where low-carbon renewables ($C_{RE} = 16.5$) are used. The snapshot for such implementation is shown in Figure 3.3, with all steps identical to those for the case with carbon-neutral resource. As shown, the minimum renewable and minimum excess energy targets are determined as 1,042,735 and 642,735 TJ, given in the first and last entries in column 10 (Figure 3.3), respectively.

We once again notice that the targets determined through the algebraic technique are more accurate than those determined using the graphical technique, i.e., EPPD (Figure 2.6).

C_t (t/TJ)	ΔC (t/TJ)	$\Sigma_j F_{Dj}$ (TJ)	$\Sigma_i F_{Si}$ (TJ)	$F_{Net, t}$ (TJ)	Infeasible cascade Cum. $F_{Net, t}$ (TJ)	ΔE_t (t)	Cum. ΔE_t (t)	$F_{CS, t}$ (TJ)	Feasible cascade Cum. $F_{Net, t}$ (TJ)	ΔE_t (t)	Cum. ΔE_t (t)
16.5				0	0				1,042,735	Minimum renewable target	
	3.5				0	0			1,042,735	3,649,573	
20		1,000,000		-1,000,000			0	0			3,649,573
	30				-1,000,000	-30,000,000			42,735	1,282,051	
50		400,000		-400,000			-30,000,000	-895,522			4,931,624
	5				-1,400,000	-7,000,000			-357,265	-1,786,325	
55			200,000	200,000			-37,000,000	-961,039			3,145,299
	20				-1,200,000	-24,000,000			-157,265	-3,145,299	
75			800,000	800,000			-61,000,000	-1,042,735			0 (Pinch)
	25				-400,000	-10,000,000			642,735	16,068,376	
100		600,000		-600,000			-71,000,000	-850,299			16,068,376
	5				-1,000,000	-5,000,000			42,735	213,675	
105			600,000	600,000			-76,000,000	-858,757			16,282,051
	4895				-400,000	-1,958,000,000			642,735	3,146,188,034	
5000							-2,034,000,000	-408,147		Minimum excess energy	3,162,470,085

FIGURE 3.3
Algebraic targeting technique with MS Excel (with low-carbon resource).

3.2 Automated Targeting Model

The algebraic targeting technique may be conveniently extended into an optimization framework, which is called the ATM (Lee et al., 2007). The basic structure of ATM is shown in Table 3.3. To perform targeting using the ATM, steps i–iii of the algebraic targeting technique (see Section 3.1) are followed. Doing this allows the energy demands and sources to be located at their respective carbon intensity level (C_k), in the first three columns of the basic structure (Table 3.3). The net energy for each level k ($F_{\text{Net}, k}$) is then calculated with Equation 3.1 (step iii). We next proceed to formulate the constraints in ATM to construct the energy and emission load cascades.

Constraint in Equation 3.7 describes the energy cascade across all carbon intensity levels. As shown, the *residual energy* from level k (δ_k) is given by the summation of residual energy from previous level (δ_{k-1}) with the net energy at level k. Note that this is conceptually similar to step iv of the algebraic targeting technique.

$$\delta_k = \delta_{k-1} + F_{\text{Net},k} \quad \forall k \tag{3.7}$$

Constraint in Equation 3.8 indicates how the *residual emission load* (ε_k) in the emission load cascade is calculated. For the first intensity level, residual emission load is set to zero, i.e., $\varepsilon_1 = 0$. On the other hand, the residual loads in other levels are contributed by the residual load from previous level (ε_{k-1}), as well as that by the residual energy within each interval; the latter is given by the product of the residual energy from level k (δ_k) and the difference between two adjacent quality levels ($\Delta C_k = C_{k+1} - C_k$). Note that this constraint is conceptually similar to Equation 3.3.

TABLE 3.3

Basic Structure of ATM

C_k	ΔC_k	$\Sigma_j F_{D,j}$	$\Sigma_i F_{S,i}$	$F_{\text{Net}, k}$	δ_k	ε_k
					$\delta_0 = F_{\text{RE}}$	
C_1		$\Sigma_j F_{D,j}$	$\Sigma_i F_{S,i}$	$F_{\text{Net},1}$	$\Downarrow$	$\varepsilon_1 = 0$
	ΔC_1				δ_1	$\Downarrow$
C_2		$\Sigma_j F_{D,j}$	$\Sigma_i F_{S,i}$	$F_{\text{Net},2}$	$\Downarrow$	ε_2
$\vdots$	ΔC_2	$\vdots$	$\vdots$	$\vdots$	δ_2	$\vdots$
$\vdots$	$\vdots$	$\vdots$	$\vdots$	$\vdots$	$\vdots$	$\vdots$
C_{n-1}	$\vdots$	$\Sigma_j F_{D,j}$	$\Sigma_i F_{S,i}$	$F_{\text{Net},2}$	$\vdots$	ε_{n-1}
	ΔC_{n-1}				$\delta_{n-1} = F_{\text{EX}}$	$\Downarrow$
C_n						ε_n

$$\varepsilon_k = \begin{cases} 0 & k = 1 \\ \varepsilon_{k-1} + \delta_{k-1}\Delta C_{k-1} & k \geq 2 \end{cases} \tag{3.8}$$

In Equation 3.9, the constraint indicates that the residual energy entering the first (δ_0) and leaving the last levels (δ_{n-1}) should take non-negative values.

$$\delta_{k-1} \geq 0 \quad k = 1, n \tag{3.9}$$

To ensure feasible load cascade, the constraint in Equation 3.10 indicates that all residual loads must take non-negative values, except the residual load at level 1 (which has been set to zero following Equation 3.8).

$$\varepsilon_k \geq 0 \quad k \geq 2 \tag{3.10}$$

The optimization objective may be set to minimize renewable energy for the system (F_{RE}), given as in Equation 3.11.

$$\text{minimize } F_{RE} \tag{3.11}$$

Alternatively, the optimization objective may also be set to minimize the overall cost for the entire system involving renewable energy r, given as in Equation 3.12.

$$\text{minimize } \Sigma_h \, F_{REh} \, C_{REh} \tag{3.12}$$

where F_{REh} and C_{REh} are the amount of renewable energy and its unit cost, respectively.

With Equations 3.7–3.12, the ATM will determine the amount of renewable energy needed for the system, along with other variables, i.e., residual energy (δ_k) and residual load at all levels (ε_k). Note that the intensity level (C_k) as well as energy demands $(F_{D,j})$ and sources $(F_{S,i})$ are parameters. Hence, the ATM formulation is a *linear program* (LP), which may be solved to obtain a global solution (if one exists) using a commercial optimization software. The following examples show the implementation of ATM in MS Excel Solver and LINGO.

Example 3.3 ATM with Spreadsheet for Classical Example 🖥

Example 3.2 (Tan and Foo, 2007) is revisited here to demonstrate how the ATM is implemented using MS Excel Solver.[2] A basic structure of ATM is first set up for the example, given as in Table 3.4. Note that the first four columns are essentially the same as those in the algebraic targeting technique (e.g., see Table 3.2). The residual energy (for energy cascade) and

---

[2] MS Excel Solver may be activated with the following path: File/Options/Add-ins… Manage/ Excel Add-ins/Go…

TABLE 3.4

Basic Structure of ATM for Example 3.3

C_k (t/TJ)	ΔC_k (t/TJ)	$\Sigma_j F_{D,j}$ (TJ)	$\Sigma_i F_{S,i}$ (TJ)	$F_{Net,k}$ (TJ)	δ_k (TJ)	ε_k (t)
					$\delta_0 = F_{RE}$	
0				0	⇓	$\varepsilon_1 = 0$
	20				δ_1	⇓
20		1,000,000		−1,000,000	⇓	ε_2
	30				δ_2	⇓
50		400,000		−400,000	⇓	ε_3
	5				δ_3	⇓
55			200,000	200,000	⇓	ε_4
	20				δ_4	⇓
75			800,000	800,000	⇓	ε_5
	25				δ_5	⇓
100		600,000		−600,000	⇓	ε_6
	5				δ_6	⇓
105			600,000	600,000	⇓	ε_7
	4,895				$\delta_7 = F_{EX}$	⇓
5,000						ε_8

residual loads (for emission load cascade) in the last two columns are to be determined with the ATM via Excel Solver. The latter has the setup as shown in Figure 3.4, where the non-negative constraints (Equations 3.9 and 3.10) are incorporated. Since the ATM is an LP, the Simplex LP (solver) has been chosen.

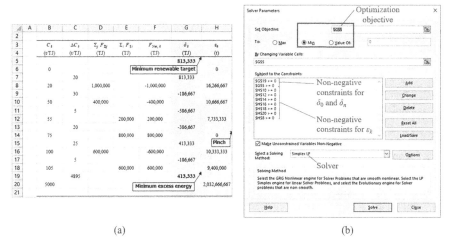

(a)   (b)

FIGURE 3.4

ATM for Example 3.3: (a) setup in spreadsheet and (b) setup in Solver.

The results of ATM in Figure 3.4 show that the minimum renewable (F_{RE}) and minimum excess energy targets (F_{EX}) are determined as 813,333 and 413,333 TJ, respectively, identical with those determined using the algebraic technique in Example 3.2 (Figure 3.2).

Example 3.4 ATM with LINGO for Classical Example (Tan and Foo, 2007) 🖥

Example 3.3 is revisited here, where the ATM is solved using a commercial optimization software, i.e., LINGO v16.[3] For this case, two renewable energy resources are available, each with their CO_2 intensity and unit cost (Chandrayan and Bandyopadhyay, 2014):

    i. Hydropower, with zero CO_2 intensity ($C_{RE1} = 0$) and unit cost (CT_{RE1}) of 28,000 \$/TJ
    ii. Biodiesel, with CO_2 intensity (C_{RE2}) of 16.5 t/TJ and unit cost (CT_{RE2}) of 31,000 \$/TJ

The objective is to determine which renewable energy is to be used for a minimum cost solution. Hence, the objective in Equation 3.12 is to be used.

The basic structure of ATM is similar to that in Example 3.3. However, note that a new intensity level (16.5 t/TJ) has been added, with biodiesel placed in the source column (see Table 3.5).

The ATM is coded in LINGO following the basic structure in Table 3.5:

```
Model:
!Objective;
min = FRE1*CTRE1 + FRE2*CTRE2;
CTRE1 = 28000;
CTRE2 = 31000;

!Constraints for residual energy (energy cascade);
D0 = FRE1;
D1 = D0;
D2 = D1 + FRE2;
D3 = D2 - 1000000;
D4 = D3 - 400000;
D5 = D4 + 200000;
D6 = D5 + 800000;
D7 = D6 - 600000;
D8 = D7 + 600000;
D8 = FEX;

!Constraints for residual loads (emission load cascade);
E1 = 0;
E2 = E1 + D1*16.5/1000;
E3 = E2 + D2*3.5/1000;
E4 = E3 + D3*30/1000;
E5 = E4 + D4*5/1000;
E6 = E5 + D5*20/1000;
```

---

[3] A free demo version of the software may be downloaded from www.lindo.com.

```
E7 = E6 + D6*25/1000;
E8 = E7 + D7*5/1000;
E9 = E8 + D8*4895/1000;

!Non-negative constraints for residual energy;
D0 >= 0;
D8 >= 0;

!Non-negative constraints for residual emission loads;
E2 >= 0; E3 >= 0; E4 >= 0; E5 >= 0;
E6 >= 0; E7 >= 0; E8 >= 0; E9 >= 0;

!Residual energy may take negative values;
@Free(d1); @Free(d2); @Free(d3); @Free(d4);
@Free(d5); @Free(d6); @Free(d7);


END
```

Note that the last set of constraints (residual energy of all levels may take negative values) is specific for LINGO, when the model is solved in LINGO, as LINGO assumes all variables have non-negative values by default. In other words, this set of constraints may be omitted when other optimization software are used.

TABLE 3.5

Basic Structure of ATM for Example 3.4

C_k (t/TJ)	ΔC_k (t/TJ)	$\Sigma_j F_{D,j}$ (TJ)	$\Sigma_i F_{S,i}$ (TJ)	$F_{Net,k}$ (TJ)	δ_k (TJ)	ε_k (×1,000 t)
					$\delta_0 = F_{RE1}$	
0				0	$\Downarrow$	$\varepsilon_1 = 0$
	16.5				δ_1	$\Downarrow$
16.5			F_{RE2}	F_{RE2}	$\Downarrow$	ε_2
	3.5				δ_2	$\Downarrow$
20		1,000,000		–1,000,000	$\Downarrow$	ε_3
	30				δ_3	$\Downarrow$
50		400,000		–400,000	$\Downarrow$	ε_4
	5				δ_4	$\Downarrow$
55			200,000	200,000	$\Downarrow$	ε_5
	20				δ_5	$\Downarrow$
75			800,000	800,000	$\Downarrow$	ε_6
	25				δ_6	$\Downarrow$
100		600,000		–600,000	$\Downarrow$	ε_7
	5				δ_7	$\Downarrow$
105			600,000	600,000	$\Downarrow$	ε_8
	4,895				$\delta_8 = F_{EX}$	$\Downarrow$
5,000						ε_9

The results of LINGO are shown as follows:

```
Global optimal solution found.
Objective value:                0.2277333E+11
Infeasibilities:                0.000000
Total solver iterations:               7
Elapsed runtime seconds:            0.18

Model Class:                           LP

Total variables:           20
Nonlinear variables:        0
Integer variables:          0

Total constraints:         29
Nonlinear constraints:      0

Total nonzeros:            56
Nonlinear nonzeros:         0
```

Variable	Value
FRE1	813333.3
CTRE1	28000.00
FRE2	0.000000
CTRE2	31000.00
D0	813333.3
D1	813333.3
D2	813333.3
D3	-186666.7
D4	-586666.7
D5	-386666.7
D6	413333.3
D7	-186666.7
D8	413333.3
FEX	413333.3
E1	0.000000
E2	13420.00
E3	16266.67
E4	10666.67
E5	7733.333
E6	0.000000
E7	10333.33
E8	9400.000
E9	2032667.

The solution from LINGO indicates that the minimum cost for the problem is determined as \$22,773 million, with ATM result shown in Table 3.6. For this case, only one renewable energy, i.e., hydropower is to be used, with minimum amount of 813,333 TJ (F_{RE1}), while the excess energy targets are determined as 413,333 TJ (F_{EX}); these are identical with those determined in Examples 3.2 and 3.3 (minimum amount solution). Besides, the pinch point occurs at residual load at level 6 (ε_6), where zero load is observed.

TABLE 3.6

Results for Example 3.4

C_k (t/TJ)	ΔC_k (t/TJ)	$\Sigma_j F_{D,j}$ (TJ)	$\Sigma_i F_{S,i}$ (TJ)	$F_{Net,k}$ (TJ)	δ_k (TJ)	ε_k (×1,000 t)
					$\delta_0 = F_{RE1} = 813,333$	
0				0	⇓	$\varepsilon_1 = 0$
	16.5				$\delta_1 = 813,333$	⇓
16.5		$F_{RE2} = 0$	$F_{RE2} = 0$		⇓	$\varepsilon_2 = 13,420$
	3.5				$\delta_2 = 813,333$	⇓
20		1,000,000		−1,000,000	⇓	$\varepsilon_3 = 16,266$
	30				$\delta_3 = -186,666$	⇓
50		400,000		−400,000	⇓	$\varepsilon_4 = 10,666$
	5				$\delta_4 = -586,666$	⇓
55			200,000	200,000	⇓	$\varepsilon_5 = 7,733$
	20				$\delta_5 = -386,666$	⇓
75			800,000	800,000	⇓	$\varepsilon_6 = 0$ (pinch)
	25				$\delta_6 = 413,333$	⇓
100		600,000		−600,000	⇓	$\varepsilon_7 = 10,333$
	5				$\delta_7 = -186,666$	⇓
105			600,000	600,000	⇓	$\varepsilon_8 = 9,400$
	4,895				$\delta_8 = 413,333$	⇓
5,000					(F_{EX})	$\varepsilon_9 = 2,032,667$

We may repeat the ATM by changing the unit price for biodiesel, in order to determine the effect of biodiesel price on the overall cost. The summary is shown in Figure 3.5, which shows that only when the biodiesel price is lower than 21,000 $/TJ, biodiesel will be considered for use. In other words, hydropower will be utilized, with the overall cost being at $ 22,773 million when the biodiesel price is higher than 21,000 $/TJ. Besides, note that higher amount of biodiesel (1,042,735 TJ) will have to be used as compared to that for hydropower (813,333 TJ), even though the overall cost is lower. These insights are very useful for practical energy planning scenario.

3.3 Conclusions

The algebraic targeting technique and ATM presented in this chapter serve as complementary tools for the graphical technique in Chapter 3. In particular, the algebraic targeting technique may be used to determine accurate targets for a carbon-constrained energy planning problem, and is

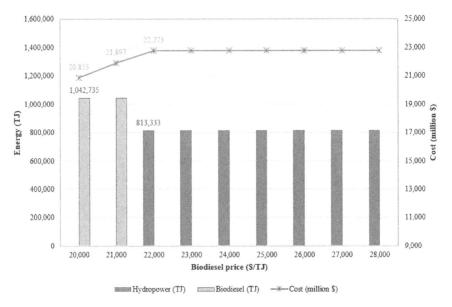

FIGURE 3.5
Effect of energy and overall cost versus biodiesel price.[4]

conveniently implemented using a spreadsheet. The ATM, on the other hand, is an optimization-based algorithm that allows the targeting steps to be performed in a commercial optimization software.

Further Reading

The algebraic targeting technique and ATM presented in this chapter may also be extended to other environmental footprints, such as water (Tan et al., 2009) and land footprints (Foo et al., 2008). Such extensions have been demonstrated for graphical technique in Chapter 2. The algebraic targeting technique can also be revised to cater multiple renewable energy resources, where *priority cost* may be used to determine the sequence of renewable energy sources to be used (see Chapter 7 for the case study in India). Besides, the ATM may be extended into *segregated targeting* where resources are dedicated for a specific application (Lee et al., 2009).

---

[4] Try to repeat this example by solving the ATM using MS Excel Solver.

References

Chandrayan, A., Bandyopadhyay, S. 2014. Cost optimal segregated targeting for resource allocation networks. *Clean Technologies and Environmental Policy*, 16, 455–465.

Foo, D. C. Y., Tan, R. R., Ng, D. K. S. 2008. Carbon and footprint-constrained energy sector planning using cascade analysis technique. *Energy*, 33(10), 1480–1488.

Lee, S. C., Ng, D. K. S., Foo, D. C. Y., Tan, R. R. 2009. Extended Pinch Targeting Techniques for Carbon-Constrained Energy Sector Planning. *Applied Energy*, 86(1), 60–67.

Tan, R. R., Foo, D. C. Y. 2007. Pinch analysis approach to carbon-constrained energy sector planning. *Energy*, 32(8), 1422–1429.

Tan, R. R., Foo, D. C. Y., Aviso, K. B., Ng, D. K. S. 2009. The use of graphical pinch analysis for visualizing water footprint constraints in biofuel production. *Applied Energy*, 86, 605–609.

Tan, R. R., Foo, D. C. Y., 2017. Energy sector planning with footprint constraints. In Abraham, M. A. (Ed.), *Encyclopedia of Sustainable Technologies*. Elsevier, Amsterdam, pp. 231–237.

4

Pinch Analysis Techniques for Carbon Capture and Storage (CCS)

In this chapter, several pinch analysis techniques developed for the planning of *CO_2 capture and storage* (CCS) will be outlined. Such systems will be needed in the near future in order to stabilize climate to safe levels (Haszeldine et al., 2018). In common technical usage, "carbon" and "CO_2" in CCS are often used interchangeably, and a similar terminology convention will be adopted here. As the name indicates, CCS consists of two sub-systems. CO_2 capture refers to isolation of a relatively pure CO_2 stream from various industrial processes, while CO_2 storage refers to the final disposal of the previously captured CO_2 in a secure, naturally occurring geological reservoir. Geological storage sites also include formations such as saline aquifers and depleted hydrocarbon reservoirs.

In this chapter, it is also assumed that CCS pertains to CO_2 emissions from the combustion products of thermal power plants. Pinch analysis techniques that are applied to two CCS subproblems, i.e., CO_2 capture and CO_2 storage are discussed. As with earlier chapters, the main philosophy of pinch analysis is to set performance targets for these sub-systems.

4.1 Carbon Capture and Storage – An Overview

Two major classes of technologies are used for carbon capture (Davison et al., 2001). First, *pre-combustion capture* relies on gasification of fuel (e.g., coal) and using water gas shift reaction to produce a gas mixture rich in hydrogen (H_2) and CO_2. The latter is separated via compression and cooling, and the remaining H_2 is used as fuel. Alternatively, in *post-combustion capture*, CO_2 can be captured directly from combustion products using techniques such as amine absorption. In such cases, the main problem is to separate the CO_2 from the rest of the flue gas, which contains a high level of N_2 from ambient air. Other alternative schemes that do not belong to pre- or post-combustion capture categories involve combustion of fuel in the absence of nitrogen (N_2), such as *oxyfuel combustion* (combustion in a mixture of O_2 and recycled CO_2)

and *chemical looping combustion*. In general, there will be an energy penalty incurred by the CO_2 capture process. If a power plant is retrofitted to allow CO_2 capture, a fraction of its previous electricity output will be lost due to this parasitic load. Thus, the energy balance impacts of such retrofits need to be compensated for by additional power generation capacity elsewhere in the grid (Tan et al., 2009).

Once point sources of CO_2 have been retrofitted for capture, the problem arises on how to transport the CO_2 to storage sites. When multiple CO_2 sources and sinks are involved, the result is a source-sink matching problem that is structurally similar to a *resource conservation network* optimization problem (Foo, 2012). The fundamental problem in such cases is ensuring that CO_2 balance constraints are met both on an annual basis and over the entire operating lifespan of the system. Thus, for a given set of CO_2 sources and sinks, the maximum potential for sequestration can be determined at an early planning stage. Network topology constraints, such as pipeline branching or merging, can be introduced during later stages of planning.

More recently, there has been growing interest in the use of captured CO_2 for various commercial applications (e.g., as a carbon-rich process feedstock) in so-called *CO_2 capture and utilization* (CCU) systems. At the macro-scale level, CCS and CCU can be integrated into *CO_2 capture, utilization, and storage* (CCUS) systems. In such cases, CCS and CCU play complementary roles, with CCS contributing permanent carbon sequestration as its main benefit (Bruhn et al., 2016). A simplified schematic of an integrated CCUS network is shown in Figure 4.1. It can be seen that there are alternative options for reuse and final disposal or storage of captured CO_2.

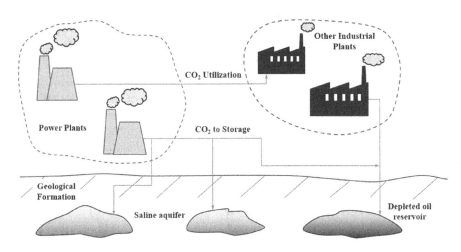

FIGURE 4.1
A macro-scale level perspective of CCUS systems.

4.2 Graphical Targeting Technique for Carbon Capture

Established techniques for carbon capture such as pre- and post-combustion systems may be applied via retrofit to existing fossil fuel-fired power plants that emit CO_2. Note that installation of these systems incurs high capital cost and partial parasitic loss of the power plant output. The latter effect implies loss of revenue from electricity sales, and thus higher generation cost. It also entails reduced thermal efficiency of the power plants.

In this section, the *carbon capture planning diagram* (CCPD) is outlined. The latter is extended from the *energy planning pinch diagram* (EPPD) introduced in Chapter 2. In particular, only the *source composite curve* of the EPPD is plotted. For a given overall emission limit of the power generation sector, the CCPD may be used to determine the minimum *extent of retrofit*, as well as the minimum amount of *compensatory power* that is required to fulfil the energy demands of the system.

Procedure for targeting using the CCPD is given as follows.

   i. Arrange the energy sources in ascending order based on their carbon intensity. Plot the source composite curve. The vertical distance of the latter corresponds to the total CO_2 emission load from the energy sources (Δm_T, Figure 4.2a).

  ii. Determine the CO_2 load removal (Δm_R) based on the *removal ratio* (*RR*) of the carbon capture system, using Equation 4.1.

$$\Delta m_R = \frac{\Delta m_T - \Delta m_L}{RR} \qquad (4.1)$$

where Δm_L is the maximum allowable CO_2 emission load. Two horizontal lines are added on the source composite curve to indicate the exact locations of Δm_R and Δm_L, respectively, as shown in Figure 4.2a.

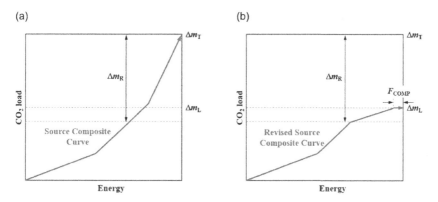

FIGURE 4.2
(a and b) CCPD for targeting carbon capture system.

iii. The minimum extent of retrofit (F_{RET}) for the carbon capture system is then calculated from Equation 4.2 that follows.

$$F_{RET} = \sum_i F_{RET,i} = \sum_i \frac{\Delta m_{R,i}}{C_{S,i}} \tag{4.2}$$

where $F_{RET,i}$, $\Delta m_{R,i}$, and $C_{S,i}$ are minimum extent of retrofit, CO_2 load removal, and carbon intensity of source i, respectively.

iv. The compensatory power (F_{COMP}) of the carbon capture system is next determined using Equation 4.3.

$$F_{COMP} = F_{RET} X \tag{4.3}$$

where X is the *parasitic energy loss ratio* of the carbon capture system. For purposes of preliminary planning, this ratio is assumed to be constant for all power plants in the system.

v. A revised CCPD may be plotted to show the source composite curve that achieves the final CO_2 emission limit (see Figure 4.2b).

Example 4.1 Graphical Targeting Technique for Carbon Capture Example (Tan et al., 2009)

This example is based on the power generation sector of the Philippines in 2007. As shown in Table 4.1, coal, natural gas, and renewables were the three major power sources for the power generation sector in the Philippines. However, due to the fossil fuel sources (i.e., coal, oil, and natural gas), the total CO_2 emission (Δm_T) of the country is estimated at 31.2 Mt/y (see last cell). A CO_2 emission limit (Δm_L) is taken as 15 Mt/y. It is assumed that the RR of the carbon capture unit as 0.85, while the parasitic energy loss ratio (X) for this case as 0.15.

Following steps i and ii of the targeting procedure, the resulting CCPD is shown in Figure 4.3. The CO_2 load removal (Δm_R) is determined using Equation 4.1 as 19.1 Mt/y (= (31.2–15)/0.85 Mt/y). This means that CO_2 load emitted from coal power plant (16.8 Mt/y, see Table 4.1) is to be completely captured, while the remaining CO_2 load 2.3 Mt/y

TABLE 4.1

Data for Example 4.1 (Tan et al., 2009)

Source i	Energy, $F_{S,i}$ (TWh/y)	Carbon Intensity, $C_{S,i}$ (Mt/TWh)	CO_2 Emission, $E_{S,i}$ (Mt/y)
Renewables	18.0	0	0
Natural gas	19.2	0.5	9.6
Oil	6.0	0.8	4.8
Coal	16.8	1.0	16.8
Total	60.0		31.2

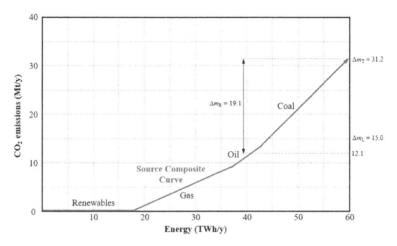

FIGURE 4.3
CCPD for Example 4.1.

(= 19.1–16.8 Mt/y) is to be captured from oil power plants. Hence, Equation 4.2 (step iii) is next followed to determine the minimum extent of retrofit for these sources, i.e., 16.8 TWh/y (= 16.8 Mt/y/1 Mt/TWh) for coal power plant, 2.9 TWh/y (= 2.3 Mt/y/0.8 Mt/TWh) for oil power plant, respectively. In other words, the overall minimum extent of retrofit (F_{RET}) can be determined as 19.7 TWh/y (= 16.8 + 2.9 TWh/y).

Next, the compensatory power (F_{COMP}) may be calculated with Equation 4.3 in step iv as 2.95 TWh/y (= 19.7 × 0.15 TWh/y). A revised CCPD is also plotted to show the final targets of the carbon capture system (Figure 4.4).

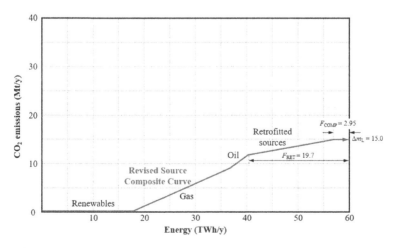

FIGURE 4.4
Revised CCPD for Example 4.1.

4.3 Automated Targeting Model (ATM) for Carbon Capture

The ATM in Chapter 3 may be extended for the targeting of carbon capture system. The basic structure of the ATM is shown in Table 4.2. Steps to perform targeting using the ATM are given as follows (Ooi et al., 2013).

i. Carbon intensity for retrofitted energy source i ($C_{R,i}$) is determined using Equation 4.4. Note that due to power losses, carbon intensities of the retrofitted sources are no longer identical as those prior to carbon capture.

$$C_{R,i} = \frac{C_i(1 - RR_i)}{1 - X_i} \quad \forall i \tag{4.4}$$

ii. Carbon intensity (C_k) for all energy sources (which include compensatory power, as well as retrofitted and non-retrofitted sources) and demand are arranged in ascending order, in column 1 (see Table 4.2). For this case, energy demand corresponds to the total requirement of the power generation sector. An arbitrary large value is added at the last level n to facilitate calculation. The difference between adjacent levels of intensity is then calculated in column 2.

iii. In column 3, the total power requirement (F_P) is located at its intensity level.

iv. The original energy output of the sources ($F_{S,i}$) and those to be retrofitted ($F_{Ret,i}$) are located at their respectively intensity levels, in columns 4 and 5, respectively. Note that due to power losses (X), the retrofitted sources will have less power output (indicated by $F_{Ret,i}(1-X_i)$). New constraints are added to ensure that the amount of

TABLE 4.2

Basic Structure of ATM for Targeting Carbon Capture System

C_k	ΔC_k	F_P	$\Sigma_i F_{S,i}$	$\Sigma_i F_{Ret,i}$	$F_{NS,k}$	$F_{Net,k}$	δ_k	ε_k
							$\delta_0 = 0$	
C_1			$\Sigma_i F_{S,i}$		$F_{NS,1}$	$F_{Net,1}$	$\Downarrow$	$\varepsilon_1 = 0$
	ΔC_1						δ_1	$\Downarrow$
C_2		F_P	$\Sigma_i F_{S,i}$	$\Sigma_i F_{Ret,i}(1-X_i)$	$F_{NS,2}$	$F_{Net,2}$	$\Downarrow$	ε_2
$\vdots$	ΔC_2		$\vdots$	$\vdots$	$\vdots$	$\vdots$	δ_2	$\vdots$
$\vdots$	$\vdots$		$\vdots$	$\vdots$	$\vdots$	$\vdots$	$\vdots$	$\vdots$
$\vdots$	$\vdots$		$\vdots$	$\vdots$	$\vdots$	$\vdots$	$\vdots$	$\vdots$
C_{n-1}	$\vdots$		$\Sigma_i F_{S,i}$	$\Sigma_i F_{Ret,i}$	$F_{NS,k}$	$F_{Net,k}$	$\Downarrow$	ε_{n-1}
	ΔC_{n-1}						δ_{n-1}	$\Downarrow$
C_n								ε_n

sources to be retrofitted will take non-negative values (Equation 4.5) and are bound to their maximum output values (Equation 4.6).

$$F_{\text{Ret},i} \geq 0 \quad \forall i \tag{4.5}$$

$$F_{\text{Ret},i} \leq F_{\text{S},i} \quad \forall i \tag{4.6}$$

v. The *net source* in column 6 ($F_{\text{NS},k}$) takes into account the net amount of energy at each intensity level, and is determined with Equation 4.7.

$$F_{\text{NS},k} = \begin{cases} F_{\text{S},i} + F_{\text{Ret},i}(1 - X_i) & \forall \text{ retrofitted source} \\ F_{\text{S},i} - F_{\text{Ret},i} & \forall i \end{cases} \tag{4.7}$$

vi. In column 7, the *net energy* for each level k ($F_{\text{Net},k}$) is determined from the difference between the total energy sources with the total power requirement, given as in Equation 4.8.

$$F_{\text{Net},k} = F_{\text{NS},k} - F_{\text{P}} \quad \forall k \tag{4.8}$$

vii. In column 8, the *cumulative energy* at interval k (δ_k) is determined using Equation 4.9, which then forms the *energy cascade*.

$$\delta_k = \begin{cases} 0 & k = 0 \\ \delta_{k-1} + F_{\text{Net},k} & k \geq 1 \end{cases} \tag{4.9}$$

viii. In column 9, the *cumulative emission load* of level k (ε_k) is calculated following Equation 4.10. This forms the *emission load cascade*.

$$\varepsilon_k = \begin{cases} 0 & k = 1 \\ \varepsilon_{k-1} + \delta_{k-1}\Delta C_{k-1} & k \geq 2 \end{cases} \tag{4.10}$$

To ensure feasible load cascade, the constraint in Equation 4.11 is added to ensure that all residual loads must take non-negative values, except the residual load at level 1 (which has been set to zero following Equation 4.10).

$$\varepsilon_k \geq 0 \quad k \geq 2 \tag{4.11}$$

ix. The objective of the problem is set to minimize the compensatory power (F_{COMP}):

$$\text{Minimize } F_{\text{COMP}} \tag{4.12}$$

The use of ATM for targeting the carbon capture system is shown next using two examples.

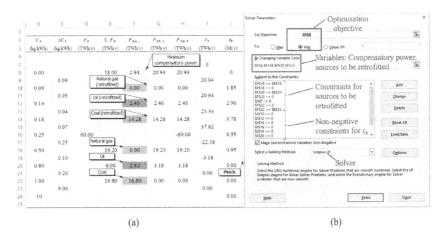

FIGURE 4.5
ATM for Example 4.2: (a) structure of ATM in Excel spreadsheet and (b) setup in Excel Solver.

Example 4.2 ATM with MS Excel (Tan et al., 2009) 🖥

Example 4.1 is revisited, to demonstrate how ATM can be solved using MS Excel Solver. The structure of ATM is set up following that in Table 4.2, and is shown in Figure 4.5a. The energy and emission load cascades in the last two columns are solved using Excel Solver. The latter has the setup as shown in Figure 4.5b, where constraints (Equations 4.5, 4.6, and 4.11) are incorporated. Since the ATM is a linear program (LP), Simplex LP is chosen as the solution algorithm. Any solution found for this LP is a global optimum.

The results of ATM are shown in Figure 4.5a. As shown, 16.8 and 2.82 TWh/y of power from coal and oil power plants are to be retrofitted with carbon capture system (see the last two entries in column F). Due to power losses, their ultimate outputs were reduced to 14.28 and 2.4 TWh/y (third and fourth entries in column F), respectively. The minimum compensatory power is determined as 2.94 TWh/y, as shown in Figure 4.5a. These targets are identical to those reported in Example 4.1.

Example 4.3 ATM with LINGO (Tan et al., 2010) 🖥

An example from Tan et al. (2010) is used to illustrate the ATM for carbon capture system. Table 4.3 shows the data of ten power plants, which generate power from different fuel types, such as coal (plants 1–5), natural gas (plants 6–8), and oil (plants 9–10). The average CO_2 intensity may be calculated by dividing the total CO_2 emission (2,490 t/h) by its energy output (3,100 MW), i.e., 0.803 t/MWh. Carbon capture is to be considered for this case, with an aim to reduce the CO_2 emission (1,100 t/h). In other words, the average CO_2 intensity is to be reduced to 0.355 t/MWh (= 1,100 t/h /3,100 MW). For this case, the carbon capture facility is assumed to have an RR of 0.9, while the parasitic energy loss ratio (X) for this case is taken as 0.25 (Tan et al., 2010). It is important to determine which power

TABLE 4.3

Data for Example 4.3 (Tan et al., 2010)

Source i	Fuel Type	Energy, $F_{S,i}$ (MW)	Carbon Intensity, $C_{S,i}$ (t CO_2/MWh)	CO_2 Emission, $E_{S,i}$ (t CO_2/h)
1	Coal	200	1	200
2	Coal	250	1	250
3	Coal	150	1	150
4	Coal	600	1	600
5	Coal	500	1	500
6	Natural gas	250	0.5	125
7	Natural gas	300	0.5	150
8	Natural gas	400	0.5	200
9	Oil	200	0.7	140
10	Oil	250	0.7	175
Total		3,100		2,490

plant should be retrofitted for carbon capture. The compensatory power will come from renewable energy sources, with an assumed carbon intensity of 0.1 t CO_2/MWh (C_{REN}).

Since individual plants are to be identified, binary variables are to be used in conjunction with the ATM. This is given by Equation 4.13.

$$F_{Ret,i} = \sum_i B_{Ret,i} F_{S,i} \qquad (4.13)$$

where $B_{Ret,i}$ is the binary variable (0 or 1) that determines if a power plant will be installed with carbon capture facilities. Due to the binary variable, the model becomes a *mixed integer linear program* (MILP).

First, the basic structure of ATM is set up and given in Table 4.4.

The ATM is coded in LINGO following the basic structure in Table 4.4:

```
!Objective;
min = FCOMP;

!Constraints for residual energy (energy cascade);
D0 = 0;
D1 = D0 + FRNG* (1 - X);
D2 = D1 + FROIL*(1 - X);
D3 = D2 + FCOMP;
D4 = D3 + FRCL* (1 - X);
D5 = D4 - 3100;
D6 = D5 + FNG - FRNG;
D7 = D6 + FOIL - FROIL;
D8 = D7 + FCL - FRCL;
D8 = FEX;

X = 0.25;

!Constraints for energy sources;

FCL = 200 + 250 + 150 + 600 + 500;
```

TABLE 4.4

Basic Structure of ATM for Example 4.3

C_k	ΔC_k	F_P	$\Sigma_i F_{S,i}$	$\Sigma_i F_{Ret,i}$	$F_{NS,k}$	$F_{Net,k}$	δ_k	ε_k
							$\delta_0 = 0$	
0.067				$F_{R,NG}(1-X)$	$F_{NS,1}$	$F_{Net,1}$	$\Downarrow$	$\varepsilon_1 = 0$
	0.026						δ_1	$\Downarrow$
0.093				$F_{R,OIL}(1-X)$	$F_{NS,2}$	$F_{Net,2}$	$\Downarrow$	ε_2
	0.007						δ_2	$\Downarrow$
0.100			F_{COMP}		$F_{NS,3}$	$F_{Net,3}$	$\Downarrow$	ε_3
	0.033						δ_3	$\Downarrow$
0.133				$F_{R,CL}(1-X)$	$F_{NS,4}$	$F_{Net,4}$	$\Downarrow$	ε_4
	0.222						δ_4	$\Downarrow$
0.355		3,100			$F_{NS,5}$	$F_{Net,5}$	$\Downarrow$	ε_5
	0.145						δ_5	$\Downarrow$
0.500			F_{NG}	$F_{R,NG}$	$F_{NS,6}$	$F_{Net,6}$	$\Downarrow$	ε_6
	0.200						δ_6	$\Downarrow$
0.700			F_{OIL}	$F_{R,OIL}$	$F_{NS,7}$	$F_{Net,7}$	$\Downarrow$	ε_7
	0.300						δ_7	$\Downarrow$
1.000			F_{CL}	$F_{R,CL}$	$F_{NS,8}$	$F_{Net,8}$	$\Downarrow$	ε_8
	9.000						δ_8	$\Downarrow$
10.000								ε_9

```
FNG = 250 + 300 + 400;
FOIL = 200 + 250;

FRCL = 200*BRET1 + 250*BRET2 + 150*BRET3 + 600*BRET4 +
500*BRET5;
FRNG = 250*BRET6 + 300*BRET7 + 400*BRET8;
FROIL = 200*BRET9 + 250*BRET10;

!Binary variables;
@BIN(BRET1); @BIN(BRET2); @BIN(BRET3); @BIN(BRET4); @BIN(BRET5);
@BIN(BRET6); @BIN(BRET7); @BIN(BRET8); @BIN(BRET9); @BIN(BRET10);

!Constraints for cumulative emission load (load cascade);
E1 = 0;
E2 = E1 + D1*0.026;
E3 = E2 + D2*0.007;
E4 = E3 + D3*0.033;
E5 = E4 + D4*0.222;
E6 = E5 + D5*0.145;
E7 = E6 + D6*0.200;
E8 = E7 + D7*0.300;
E9 = E8 + D8*9.000;

!Non-negative constraints for residual energy;
D0 >= 0;
D8 >= 0;
```

```
!Non-negative constraints for cumulative emission load;
E2 >= 0; E3 >= 0; E4 >= 0; E5 >= 0;
E6 >= 0; E7 >= 0; E8 >= 0; E9 >= 0;

!Residual energy may take negative values;
@Free(d1); @Free(d2); @Free(d3); @Free(d4);
@Free(d5); @Free(d6); @Free(d7);


END
```

Note again that the last set of constraints (residual energy of all levels may take negative values) is specific for LINGO, since when the model is solved in LINGO, all variables are assumed to have non-negative values by default. The @Free command deactivates this default assumption. Thus, this set of constraints should be omitted when other optimization software are used.

The LINGO solution report is shown as follows:

```
Global optimal solution found.
Objective value:              412.5000
Objective bound:              412.5000
Infeasibilities:              0.000000
Extended solver steps:               1
Total solver iterations:            65
Elapsed runtime seconds:          0.07

Model Class:                      MILP

Total variables:              31
Nonlinear variables:           0
Integer variables:            10

Total constraints:            31
Nonlinear constraints:         0

Total nonzeros:               70
Nonlinear nonzeros:            0


              Variable            Value
                 FCOMP        412.5000
                    D0        0.000000
                    D1        0.000000
                  FRNG        0.000000
                     X       0.2500000
                    D2        150.0000
                 FROIL        200.0000
                    D3        562.5000
                    D4        1650.000
                  FRCL        1450.000
                    D5       -1450.000
                    D6       -500.0000
                   FNG        950.0000
```

D7	-250.0000
FOIL	450.0000
D8	0.000000
FCL	1700.000
FEX	0.000000
BRET1	1.000000
BRET2	0.000000
BRET3	1.000000
BRET4	1.000000
BRET5	1.000000
BRET6	0.000000
BRET7	0.000000
BRET8	0.000000
BRET9	1.000000
BRET10	0.000000
E1	0.000000
E2	0.000000
E3	1.050000
E4	19.61250
E5	385.9125
E6	175.6625
E7	75.66250
E8	0.6625000
E9	0.6625000

The global optimum solution from LINGO indicates that the minimum compensatory power (FCOMP) is determined as 412.5 MW. The results also indicate that power plants to be retrofitted with carbon capture system are mostly coal plants (plants 1, 3–5) and one oil power plant (plant 9). No gas power plant is to be retrofitted with carbon capture as the fuel is relatively clean (i.e., with low CO_2 intensity). A summary of the results is given in Table 4.5.

TABLE 4.5

Summary for Example 4.3 (Tan et al., 2010)

Source i	Fuel Type	Retrofitted	Final Output (MW)	Carbon Intensity, $C_{s,i}$ (t CO_2/MWh)	CO_2 Emission, $E_{s,i}$ (t CO_2/h)
1	Coal	Yes	150	0.133	20.0
2	Coal	No	250	1.000	250.0
3	Coal	Yes	112.5	0.133	15.0
4	Coal	Yes	450	0.133	59.9
5	Coal	Yes	375	0.133	49.9
6	Natural gas	No	250	0.500	125.0
7	Natural gas	No	300	0.500	150.0
8	Natural gas	No	400	0.500	200.0
9	Oil	Yes	150	0.093	14.0
10	Oil	No	250	0.700	175.0
	Compensatory		412.5	0.1000	41.3
Total			**3,100**		**1,099.8**

4.4 Graphical Targeting Technique for CO_2 Storage

Once CO_2 is captured from their point sources (e.g., industrial flue gases), it needs to be sent for storage in some secure natural reservoir (i.e., *sink*). The latter may include depleted oil or gas wells, inaccessible coal seams, and saline aquifers. The task is to determine the optimum matching among the CO_2 sources and sinks. Note, however, that there are several inherent limitations for the CO_2 sinks:

- All sinks have limited capacity which limits the amount of CO_2 to be stored over their entire operating life. Such limits are based on the physical size of the storage reservoir.
- All sinks have injectivity constraints which limit the annual flowrate of CO_2 injection. These limits are based on the porosity and permeability of the geological strata used for storage.
- All sinks are assumed to be immediately available at the beginning of the time planning horizon ($t = 0$) for the injection of CO_2 sources.

In this section, the *carbon storage pinch diagram* (CSPD) is introduced. The latter is conceptually similar to the EPPD introduced in Chapter 2, in which composite curves are used to determine the system targets for the matching of CO_2 sources and sinks. For this case, the system targets correspond to the *unutilized capacity* and *storage deficit* of the carbon capture system. Network topology constraints are not considered in this approach, but are assumed to come into play at a later stage of the planning process. Once such constraints are taken into account, the resulting solutions may not be able to achieve the optimum targets of unrestricted networks.

Procedure for targeting using the CSPD is given as follows (Diamante et al., 2013).

   i. All CO_2 sources and sinks are arranged respectively in descending order of their lifespan.

   ii. The energy demands are plotted in a CO_2 flow vs. CO_2 flowrate to form the *sink composite curve*, in order of descending lifespan.

   iii. The energy sources are plotted in a CO_2 flow vs. CO_2 flowrate to form the *source composite curve*, in order of descending lifespan.

   iv. The demand and source composite curves are superimposed in the same CO_2 flow vs. CO_2 flowrate diagram to form the CSPD.

   v. In cases where the source composite curve is entirely below and to the right of the sink composite curve, and provided the former has a larger CO_2 capacity (*x*-axis) than that of the source composite curve, the CSPD is considered feasible.

vi. For cases where the entire or part of the source composite curve stays to the left or above the sink composite curve, the CSPD is considered as infeasible. To obtain a feasible CSPD, the source composite curve is to be shifted horizontally to the right, until it lies entirely to the right and below the sink composite curve. The two composite curves touch each other at the *pinch point*.

For both feasible CSPDs in steps v and vi, the portion of the sink composite curve extended to the left side of the source composite curve defines the minimum unutilized capacity of the carbon capture system. On the other hand, the part of the source composite curve extended to the right of sink composite curve indicates the storage deficit of the system. In other words, larger storage capacity is necessary for the CO_2 capture task.

An example is next used to elucidate the graphical targeting tool.

Example 4.4 ATM with Graphical Technique for Literature Example (Diamante et al., 2013)

In this example, a total of four CO_2 sources are to be sent to two CO_2 sinks, with data given in Table 4.6. Not that the sources and sinks have been arranged in descending order of lifespan, given in column 3 (step i). Note also that the flowrate data in the last column is obtained by dividing the CO_2 flows (column 2) with the lifespan values (column 3).

Following steps ii–iv of the procedure, the sink and source composite curves are plotted to form the CSPD, as shown in Figure 4.6. The figure shows that the source composite curve stays to the left and above the sink composite curve; hence, the CSPD is considered as infeasible (step v).

Step vi is next followed to shift the source composite curve horizontally to the right, until it stays entirely to the right and below the sink composite curve, and touches the latter at the pinch. This forms the feasible CSPD in Figure 4.7. The latter shows that the minimum unutilized

TABLE 4.6

Data for Example 4.4: (a) CO_2 Sinks and (b) CO_2 Sources (Diamante et al., 2013)

Sink j	CO_2 Flow, $m_{SK,j}$ (Mt)	Lifespan, $y_{SK,j}$ (y)	CO_2 Flowrate, $F_{SK,j}$ (Mt/y)
(a)			
SK1	750	50	15
SK2	250	10	25
Source i	CO_2 Flow, $m_{SR,i}$ (Mt)	Lifespan, $y_{SR,i}$ (y)	CO_2 Flowrate, $F_{SR,i}$ (Mt/y)
(b)			
SR1	200	40	5
SR2	300	30	10
SR3	200	25	8
SR4	100	20	5

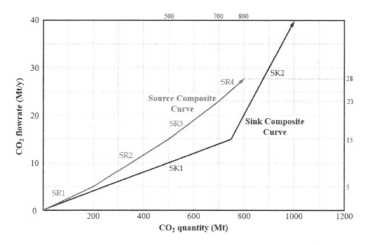

FIGURE 4.6
An infeasible CSPD for Example 4.4.

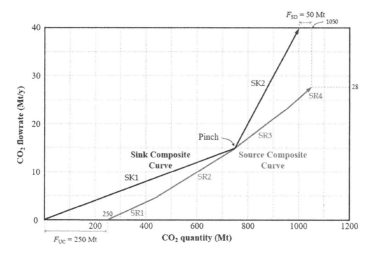

FIGURE 4.7
A feasible CSPD for Example 4.4 (Diamante et al., 2013).

capacity (F_{UC}) of the carbon capture system is identified as 250 Mt, indicated by the opening on the left side of the source composite curve. On the other hand, Figure 4.7 also shows that the storage deficit (F_{SD}) of the system is determined as 50 Mt, given by the section of the source composite curve extended to the right of the sink composite curve. Hence, larger storage capacity is necessary to fulfil the CO_2 storage requirement. In practice, it may be necessary to export this excess CO_2 outside of the system (assuming potential sinks are available in adjacent regions). An alternative interpretation is to forego the capture of CO_2 at the source that extends beyond the available storage sites.

4.5 Conclusions

Different process integration tools were presented in this chapter for the two important elements of CCS, i.e., carbon capture and carbon storage. Graphical, algebraic, and mathematical programming variants have been discussed. These tools are based on the main philosophy of process integration, i.e., *target* setting, which enables transparent high-level planning of CCS systems for effective decision support. These targets provide planners with information on physically feasible and optimal levels of CO_2 sequestration for a given system.

Further Reading

The ATM for carbon capture presented in this chapter can be extended for multi-period consideration (Ooi et al., 2014). A country case study for this scenario is shown in Chapter 8, where simultaneous planning of power generation and desalination in United Arab Emirates (UAE) is examined. For the carbon storage system, one may also consider the availability, i.e., start and end times of the CO_2 storage reservoirs. This calls for the use of another targeting tool along with inspection (Thengane et al., 2019). The methodology may also be extended to alternative carbon sequestration techniques, such as biochar-based carbon management networks (Tan et al., 2018).

References

Bruhn, T., Naims, H., Olfe-Kräutlein, B. 2016. Separating the debate on CO_2 utilisation from carbon capture and storage. *Environmental Science and Policy*, 60, 38–43.

Davison, J., Freund, P., Smith, A. 2001. *Putting Carbon Back into the Ground. International Energy Agency Greenhouse Gas R&D Programme*, Cheltenham.

Foo, D. C. Y. 2012. *Process Integration for Resource Conservation*. CRC Press, Boca Raton, FL.

Haszeldine, R. S., Flude, S., Johnson, G., Scott, V. 2018. Negative emissions technologies and carbon capture and storage to achieve the Paris Agreement commitments. *Philosophical Transactions of the Royal Society A: Mathematical, Physical and Engineering Sciences*. doi: 0.1098/rsta.2016.0447.

Ooi, R. E. H., Foo, D. C. Y., Tan, R. R., Ng, D. K. S., Smith, R. 2013. Carbon Constrained Energy Planning (CCEP) for sustainable power generation sector with automated targeting model. *Industrial & Engineering Chemistry Research*, 52, 9889–9896.

Ooi, R. E. H., Foo, D. C. Y., Tan, R. R. 2014. Targeting for carbon sequestration retrofit planning in the power generation sector for multi-period problems. *Applied Energy*, 113, 477–487.

Tan, R. R., Ng, D. K. S., Foo, D. C. Y. 2009. Pinch analysis approach to carbon-constrained planning for sustainable power generation. *Journal of Cleaner Production*, 17(10), 940–944.

Tan, R. R., Ng, D. K. S., Foo, D. C. Y., Aviso, K. B. 2010. Crisp and Fuzzy integer programming models for optimal carbon sequestration retrofit in the power sector. *Chemical Engineering Research and Design*, 88(12), 1580–1588.

Tan, R. R., Bandyopadhyay, S., Foo, D. C. Y. 2018. Graphical pinch analysis for planning biochar-based carbon management networks. *Process Integration and Optimization for Sustainability*, 2, 159–168.

Thengane, S. K., Tan, R. R., Foo, D. C. Y., Bandyopadhyay, S., 2019. A pinch-based approach for targeting of Carbon Capture, Utilization and Storage (CCUS) Systems. *Industrial & Engineering Chemistry Research*, 58, 3188–3198.

5

Superstructure-Based
Mathematical Programming

This chapter outlines a commonly used mathematical programming technique, i.e., *superstructure* approach for energy planning problems. In the general context of process integration literature, and particularly heat integration, pinch analysis and superstructure-based optimization were originally seen as competing schools of thought; more recently, there has been a shift towards their use as complementary techniques with their respective advantages and disadvantages (Klemeš and Kravanja, 2013).

The superstructure approach serves as a complementary tool for pinch analysis techniques outlined in Chapters 2–4. A superstructure is a network that contains all possible structural configurations that may occur in the system. The optimization takes place in a single step, without prior knowledge of the optimal target; instead, an optimal network is immediately generated. Both network topology and stream flowrates are determined when the model is solved. This feature is useful for handling large, complex problems. It also allows inclusion of case-specific constraints that are unique to specific problems. On the other hand, the automated characteristic of such models also makes them less transparent to users; thus, problem analysis is not as straightforward with superstructure models as it is with pinch analysis. Engineering insights to support decision-making emerge more readily from the latter approach.

In carbon-constrained energy planning problem, it provides detailed analysis on how energy sources may be matched with the sinks. For the planning of carbon capture and storage (CCS), it provides the detailed output on which CO_2 point source is to be retrofitted for capture.

5.1 General Linear Programming (LP) Model for Carbon-Constrained Energy Planning Problem

The mathematical formulation for carbon-constrained energy planning takes the form as a source-demand matching problem, as shown in the superstructure representation as in Figure 5.1 (Tan and Foo, 2007, Pękala et al., 2010). It is interesting to note that it resembles the sink–source matching problem in *material resource conservation networks* (Foo, 2012). As shown

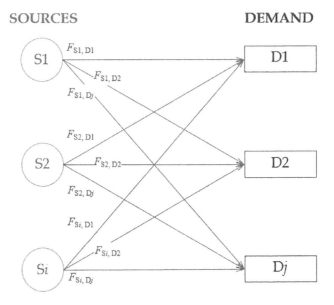

FIGURE 5.1
Superstructure representation for a source-demand matching problem (adapted from Foo, 2012).

in Figure 5.1, all energy sources are linked to the demands with a potential connection. The actual existence of the connections will be determined by the mathematical model.

The optimization objective may be set to minimize total renewable energy (F_{RE}) of the system; the latter is given as the summation of renewable energy that is allocated to the individual demand j ($F_{RE,j}$) given as in Equation 5.1.

$$\text{minimize } F_{RE} = \sum_j F_{RE,j} \tag{5.1}$$

Subject to:

$$F_{RE,j} + F_{i,j} = F_{D,j} \quad \forall j \tag{5.2}$$

$$F_{RE,j}C_{RE} + \sum_i F_{i,j}C_{S,i} \leq F_{D,j}C_{D,j} \quad \forall j \tag{5.3}$$

$$\sum_j F_{i,j} + F_{EX,i} = F_{S,i} \quad \forall i \tag{5.4}$$

$$F_{EX,i} \geq 0 \quad \forall i \tag{5.5}$$

$$F_{RE,j} \geq 0 \quad \forall j \tag{5.6}$$

$$F_{i,j} \geq 0 \quad \forall i \; \forall j \tag{5.7}$$

where variable $F_{i,j}$ is the allocation of energy from source i to demand j, variable $F_{EX,i}$ is the excess energy from source i, parameter $F_{S,i}$ is the amount of energy available from source i, parameter $F_{D,j}$ is the energy requirement of demand j, parameter C_{RE} is the carbon intensity of the renewable energy resource, parameter $C_{S,i}$ is the carbon intensity of source i, and parameter $C_{D,j}$ is the carbon intensity limit of demand j.

In this formulation, the energy and carbon load balances for the demands are given by Equations 5.2 and 5.3, respectively. Note that C_{RE} assumes a value of zero for a carbon-neutral resource. Note also that it can even take a negative value if *negative emissions technologies* (NETs, e.g., bioenergy with CCS) are used in the system (McLaren, 2012). The energy balance for the sources is given by Equation 5.4. All variables in the system are assumed to be non-negative (Equations 5.5–5.7). This formulation is a linear program (LP), where global solution is guaranteed, if solution exists.

Unlike pinch analysis approaches, the sources and demands may be included in the model in any arbitrary order, without having to consider specific sequences based on carbon intensity. The same LP model also applies to any sustainability index for the energy sources, e.g., land and water footprint.[1]

Example 5.1 Superstructural Model for Classical Example (Tan and Foo, 2007) 🖳

The classical example (Tan and Foo, 2007) is revisited, with data given in Table 2.2. In this case, three energy sources may be utilized for three energy demands, which is supplemented by carbon-neutral (i.e., $C_{RE} = 0$) and low-carbon fuel renewables (i.e., $C_{RE} = 16.5$ t/TJ). The objective is to minimize the renewables.

The problem is solved using MS Excel. The superstructure is set up in the main Excel sheet (Figure 5.2a), while the constraints are set up in Solver (Figure 5.2b). Note that the last two rows of the superstructure contain the total energy and CO_2 load allocated from all the sources to the sinks, respectively.

For the case with carbon-neutral, the minimum renewable and minimum excess energy are determined as 813,333 and 413,333 TJ, respectively (see Figure 5.3), which are identical with those determined using the algebraic technique in Example 3.2.

Similarly, for the case with low-carbon renewables ($C_{RE} = 16.5$), the minimum renewable and minimum excess energy are determined as 1,042,735 and 642,735 TJ (Figure 5.4), respectively. These are also identical to those reported in Example 3.2.

---

[1] See Chapter 2 for detailed discussion.

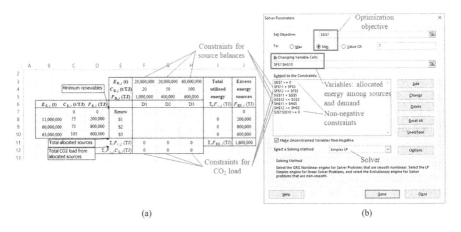

(a)   (b)

FIGURE 5.2
Configuration for Example 5.1: (a) superstructure in Excel spreadsheet and (b) setup in Excel Solver.

	$E_{D,j}$ (t)	20,000,000	20,000,000	60,000,000	Total	Excess
Minimum renewables	$C_{D,j}$ (t/TJ)	20	50	100	utilised	energy
	$F_{D,j}$ (TJ)	1,000,000	400,000	600,000	energy	sources
$E_{S,i}$ (t) $C_{S,i}$ (t/TJ) $F_{S,i}$ (TJ)		D1	D2	D3	$\Sigma_j F_{i,j}$ (TJ)	$F_{EX,i}$ (TJ)
0 813,333	Renew	680,000	133,333	0		0
11,000,000 55 200,000	S1	200,000	0	0	200,000	0
60,000,000 75 800,000	S2	120,000	266,667	100,000	486,667	313,333
63,000,000 105 600,000	S3	0	0	500,000	500,000	100,000
	$\Sigma_i F_{i,j}$ (TJ)	1,000,000	400,000	600,000	$\Sigma_i F_{EX,i}$(TJ)	413,333
	$\Sigma_i F_{i,j} C_{S,i}$(TJ)	20,000,000	20,000,000	60,000,000		

FIGURE 5.3
Solution for Example 5.1 with carbon-neutral resource.

	$E_{D,j}$ (t)	20,000,000	20,000,000	60,000,000	Total	Excess
Minimum renewables	$C_{D,j}$ (t/TJ)	20	50	100	utilised	energy
	$F_{D,j}$ (TJ)	1,000,000	400,000	600,000	energy	sources
$E_{S,i}$ (t) $C_{S,i}$ (t/TJ) $F_{S,i}$ (TJ)		D1	D2	D3	$\Sigma_j F_{i,j}$ (TJ)	$F_{EX,i}$ (TJ)
16.5 1,042,735	Renew	909,091	133,644	0		0
11,000,000 55 200,000	S1	90,909	109,091	0	200,000	0
60,000,000 75 800,000	S2	0	157,265	100,000	257,265	542,735
63,000,000 105 600,000	S3	0	0	500,000	500,000	100,000
	$\Sigma_i F_{i,j}$ (TJ)	1,000,000	400,000	600,000	$\Sigma_i F_{EX,i}$(TJ)	642,735
	$\Sigma_i F_{i,j} C_{S,i}$(TJ)	20,000,000	20,000,000	60,000,000		

FIGURE 5.4
Solution for Example 5.1 with low-carbon resource.

5.2 General LP Model for Carbon Capture
System at Sectoral Level

The LP model for carbon capture is an extension of that for carbon-constrained energy planning problem. It takes to following form, with superstructure shown in Figure 5.5. The latter is structurally similar to that in Figure 5.1, where all energy sources are linked to the demands directly, and through the carbon capture facilities ($R = 1, 2, r$). In this generic model, each type of energy sources may be installed with a single type of carbon capture facility of given performance specification (e.g., removal ratio, power losses).[2]

The optimization objective may be set to minimize compensatory power (F_{COMP}) of the system, given as in Equation 5.8.

$$\text{minimize } F_{\text{COMP}} = \sum_j F_{\text{COMP},j} \tag{5.8}$$

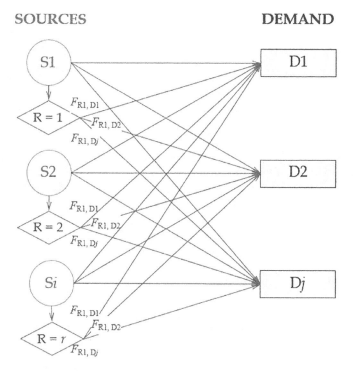

FIGURE 5.5
Supertructure for carbon capture system.

---
[2] See Chapter 4 for detailed discussion.

Subject to:

$$F_{COMP,j} + F_{i,j} + \sum_r F_{Ret,r,j} = F_{D,j} \quad \forall j \tag{5.9}$$

$$F_{COMP,j} C_{RE} + \sum_i F_{i,j} C_{S,i} + \sum_r F_{Ret,r,j} C_{Ret,r} \leq F_{D,j} C_{D,j} \quad \forall j \tag{5.10}$$

$$\sum_j F_{i,j} + \sum_r F_{S,i,r} + F_{EX,i} = F_{S,i} \quad \forall i \tag{5.11}$$

$$F_{S,i,r}(1 - X_r) = F_{Ret,r,j} \quad \forall r; r = i \tag{5.12}$$

$$C_{Ret,r}(1 - X_r) = C_{S,i}(1 - RR_r) \quad \forall r \tag{5.13}$$

$$F_{S,r} \geq 0; F_{Ret,r} \geq 0 \quad \forall r \tag{5.14}$$

$$F_{EX,i} \geq 0 \quad \forall i \tag{5.15}$$

$$F_{RE,j} \geq 0 \quad \forall j \tag{5.16}$$

$$F_{i,j} \geq 0 \quad \forall i \; \forall j \tag{5.17}$$

where variable $F_{S,i,r}$ is the energy of source i to be retrofitted with capture technique r, variable $F_{Ret,r,j}$ is the energy to be sent to demand j from retrofit technique r, parameters X_r and RR_r are the ratios of power loss and CO_2 removal of the carbon capture technique r, and C_{RE} is the CO_2 intensity of compensatory power, which is assumed to be originated from renewable energy.

Note that the first three constraints (Equations 5.9–5.11) are the revised form of those for carbon-constrained energy planning problem (Equations 5.2–5.4). Constraints in Equations 5.12 and 5.13 give the relationship of the input and output of the carbon capture technique r. All variables for the system are assumed to be non-negative (Equations 5.15–5.17).

Example 5.2 Superstructural Technique for Carbon Capture Example (Tan et al., 2009) 💻

The Philippines power generation sector example (Example 4.1) is revisited, with data given in Table 4.1. For this case, coal, oil, natural gas, and renewables are used for power generation sector in the Philippines (a total of 60 TWh/y). The total CO_2 emission of the country is estimated at 31.2 Mt/y (see the last cell in Table 4.1). Assuming a CO_2 emission limit (m_L) is to be set at 15 Mt/y, the intensity of the power generation sector is to be reduced from 0.52 Mt/TWh (= 31.2/60 Mt/TWh) to 0.25 Mt/TWh. To achieve this, carbon capture system[3] is to be used. It is assumed that only

---

[3] See Section 4.1 for detailed discussion on carbon capture and carbon storage systems.

one type of carbon capture technique is used, with *removal ratio (RR)* of 0.85, while its parasitic energy loss ratio (X) is taken as 0.15.

The superstructure is configured in the Excel spreadsheet and the various constraints are set up accordingly in its Solver, as shown in Figure 5.6a and b. Note that there is only one energy demand in this case, so all retrofitted and non-retrofitted energy sources are to be sent to this demand, and the power losses due to retrofit in each energy source are to be accounted for in the spreadsheet (Figure 5.5a).

The superstructural model is solved, with the solution shown in Figure 5.7. For this case, 2.82 and 16.8 TWh/y of power from oil (S2) and coal (S3) power plants are to be retrofitted with carbon capture system. Due to power losses, their ultimate outputs were reduced to 2.4 TWh/y (R2) and 14.28 TWh/y (R3), respectively (see row 3 from top). The minimum compensatory power (F_{COMP}) is determined as 2.94 TWh/y,

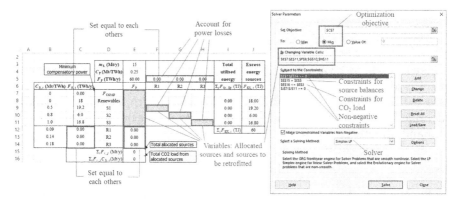

FIGURE 5.6
Configuration for Example 5.2: (a) superstructure in Excel spreadsheet and (b) setup in Excel Solver.

Minimum compensatory power		m_L (Mt/y)	15				Total utilised energy	Excess energy sources
		C_P (Mt/TWh)	0.25					
		F_P (TWh/y)	60.00	0.00	2.40	14.28		
$C_{S,i}$ (Mt/TWh)	$F_{S,i}$ (TWh/y)		F_P	R1	R2	R3	$\Sigma_j F_{Si,Dj}$ (TJ)	$F_{EX,i}$ (TJ)
0	2.94	F_{COMP}	2.94					0.00
0	18	Renewables	18.00				18.00	0.00
0.5	19.2	S1	19.20	0.00			19.20	0.00
0.8	6.0	S2	3.18		2.82		6.00	0.00
1.0	16.8	S3	0.00			16.80	16.80	0.00
0.09	0.00	R1	0.00				$\Sigma_i F_{EX,i}$ (TJ)	0
0.14	2.40	R2	2.40					
0.18	14.28	R3	14.28					
		$\Sigma_i F_{i,j}$ (Mt/y)	60	Total allocated sources				
		$\Sigma_i F_{i,j} C_{S,i}$ (Mt/y)	15	Total CO2 load from allocated sources				

FIGURE 5.7
Solution for Example 5.2.

corresponding to the total power losses. These values are identical to those reported in Example 4.3, which were determined using the pinch-based automated targeting model (ATM).

5.3 Extended Model for Carbon Capture System at Plant Level

The LP model in the earlier section is suitable for the planning of sectoral level, as it identifies which fuel sector(s) to be retrofitted with carbon capture system in order to fulfil the CO_2 emission limit. Note, however, that it does not identify which specific power plants are to be selected for carbon capture retrofit. To identify such plants, constraint in Equation 5.12 is to be modified and take the revised form in Equation 5.18.

$$F_{S,i}\left(1-X_r\right)B_r = F_{\text{Ret},r,j} \quad \forall r \, \forall j; r = i \tag{5.18}$$

where is B_r a binary variable {0,1}, which takes a value of unity when a power plant is identified for carbon capture retrofit, or zero when it is not being selected. In addition, Equation 5.19 is added to ensure that once a plant is identified for retrofit, it can no longer send its power output to the energy demand directly.

$$F_{i,j}\left(1-B_r\right)+F_{\text{RET},r,j}\,B_r = F_{S,i} \quad \forall r \, \forall j; r = i \tag{5.19}$$

Note, however, that the binary variables are multiplied by continuous variables in Equations 5.18 and 5.19, resulting in the presence of bilinear terms. This complication makes the mathematical model a *mixed integer non-linear program* (MINLP). In general, solving such models to global optimality presents computational challenges and requires specialized global optimization algorithms. For example, the commercial optimization software – LINGO uses a branch-and-bound-based global optimization toolbox to deal with such models (Gau and Schrage, 2004). Alternatively, the problem may be reformulated as a pure *integer linear program* (ILP) that can be readily solved to global optimality.[4] The alternative formulation is beyond the scope of this chapter, but the reader may refer to Tan et al. (2010) for details.

Example 5.3 Superstructural Technique for Ten Plants Case Study (Tan et al., 2010) 🖳

The ten plants case study in Example 4.3 is revisited, with data given in Table 4.3. The total power output of these power plants is 3,100 MW and their total CO_2 emission is 2,490 t/h (see Table 4.3). The CO_2 intensity is

---

[4] The LINGO code for the ILP model is available as a supplementary file on the book support website.

hence calculated as 0.803 t/MWh (= 2,490 t/h / 3,100 MW). It is desired to reduce the CO_2 intensity to 0.355 t/MWh with carbon capture technique. Power plants are to be retrofitted with carbon capture in order to achieve the objective. The *RR* of CO_2 for the selected carbon capture technique is given as 0.9, with a parasitic energy loss (*X*) of 0.25. Losses from the capture retrofit is assumed to be offset with renewable energy sources, with an average emission factor of 0.1 t CO_2/MWh. The relative cost of power generation from retrofitted plants is assumed to be 1.6, while that for compensatory power generation is 1.4. For this case, it is desired to minimize the compensatory power (with Equation 5.8), while keeping the cost rise of the power output by 30%. To account for the limitation of cost rise (i.e., cost rise should not exceed 30%), a new constraint is to be added to the model, i.e.:

$$1.4F_{COMP} + \sum_i F_{S,i} + 1.6 \sum_r F_{Ret,r} \le 1.3F_D \tag{5.20}$$

where F_D is the total power output, i.e., 3,100 MW.

The ATM is coded in LINGO as follows:

```
min = COMP;

PRICE <= 1.3*3100 ;
PRICE = 1.4*FRE + 1.0*FS + 1.6*FRET ;


!Energy balances for demands;

COMP + FS + FRET = 3100;

FS =
FS1 + FS2 + FS3 + FS4 + FS5 +
FS6 + FS7 + FS8 +
FS9 + FS10 ;

FRET =
FRET1 + FRET2 + FRET3 + FRET4 + FRET5 +
FRET6 + FRET7 + FRET8 +
FRET9 + FRET10;


!CO2 balances for demands;

COMP*0.1 + FSCO2 + RCO2 <= 3100*0.355;

FSCO2 =
(FS1 + FS2 + FS3 + FS4 + FS5)*1 +
(FS6 + FS7 + FS8)*0.5 +
(FS9 + FS10)*0.7 ;

RCO2 =
(FRET1 + FRET2 + FRET3 + FRET4 + FRET5)* CRETCOAL +
(FRET6 + FRET7 + FRET8)* CRETNG +
(FRET9 + FRET10)* CRETOIL ;
```

```
CRETCOAL = 1*(1-RR)/(1-X);
CRETNG = 0.5*(1-RR)/(1-X);
CRETOIL = 0.7*(1-RR)/(1-X);

X = 0.25 ;
RR = 0.9 ;


!Energy balances for sources (assuming no excess energy from
all sources);

FS1+ FS1RET = 200;
FS2+ FS2RET = 250;
FS3+ FS3RET = 150;
FS4+ FS4RET = 600;
FS5+ FS5RET = 500;
FS6+ FS6RET = 250;
FS7+ FS7RET = 300;
FS8+ FS8RET = 400;
FS9+ FS9RET = 200;
FS10+ FS10RET = 250;


!Plant(s) to be selected for retrofit;

200*(1 - X)* B1 = FRET1 ;
250*(1 - X)* B2 = FRET2 ;
150*(1 - X)* B3 = FRET3 ;
600*(1 - X)* B4 = FRET4 ;
500*(1 - X)* B5 = FRET5 ;
250*(1 - X)* B6 = FRET6 ;
300*(1 - X)* B7 = FRET7 ;
400*(1 - X)* B8 = FRET8 ;
200*(1 - X)* B9 = FRET9 ;
250*(1 - X)* B10 = FRET10 ;

FS1*(1-B1) + FS1RET*B1 = 200;
FS2*(1-B2) + FS2RET*B2 = 250;
FS3*(1-B3) + FS3RET*B3 = 150;
FS4*(1-B4) + FS4RET*B4 = 600;
FS5*(1-B5) + FS5RET*B5 = 500;
FS6*(1-B6) + FS6RET*B6 = 250;
FS7*(1-B7) + FS7RET*B7 = 300;
FS8*(1-B8) + FS8RET*B8 = 400;
FS9*(1-B9) + FS9RET*B9 = 200;
FS10*(1-B10) + FS10RET*B10 = 250;


!To define binary variables;

@bin(B1) ; @bin(B2) ; @bin(B3) ; @bin(B4) ; @bin(B5) ;
@bin(B6) ; @bin(B7) ; @bin(B8) ; @bin(B9) ; @bin(B10) ;


End
```

The global optimizer toolbox in LINGO needs to be used to guarantee that a global optimum is determined for this problem. This toolbox is an add-on feature that needs to be purchased and licensed separately.[5] Use of the default LINGO solver cannot guarantee that the solution found is a global optimum. The LINGO solution report is shown as follows:

```
Global optimal solution found.
Objective value:                    412.5000
Objective bound:                    412.5000
Infeasibilities:                    0.000000
Extended solver steps:                     0
Total solver iterations:                  80
Elapsed runtime seconds:                0.47

Model Class:                           MINLP

Total variables:           47
Nonlinear variables:       30
Integer variables:         10

Total constraints:         39
Nonlinear constraints:     10

Total nonzeros:           126
Nonlinear nonzeros:        30


                    Variable           Value
                        COMP        412.5000
                       PRICE        3430.000
                         FRE        0.000000
                          FS        1450.000
                        FRET        1237.500
                         FS1        0.000000
                         FS2        250.0000
                         FS3        0.000000
                         FS4        0.000000
                         FS5        0.000000
                         FS6        250.0000
                         FS7        300.0000
                         FS8        400.0000
                         FS9        0.000000
                        FS10        250.0000
                       FRET1        150.0000
                       FRET2        0.000000
                       FRET3        112.5000
                       FRET4        450.0000
                       FRET5        375.0000
                       FRET6        0.000000
                       FRET7        0.000000
                       FRET8        0.000000
                       FRET9        150.0000
```

---

[5] See Preface for the details of LINGO.

FRET10	0.000000
FSCO2	900.0000
RCO2	159.0000
CRETCOAL	0.1333333
CRETNG	0.6666667E-01
CRETOIL	0.9333333E-01
RR	0.9000000
X	0.2500000
FS1RET	200.0000
FS2RET	0.000000
FS3RET	150.0000
FS4RET	600.0000
FS5RET	500.0000
FS6RET	0.000000
FS7RET	0.000000
FS8RET	0.000000
FS9RET	200.0000
FS10RET	0.000000
B1	1.000000
B2	0.000000
B3	1.000000
B4	1.000000
B5	1.000000
B6	0.000000
B7	0.000000
B8	0.000000
B9	1.000000
B10	0.000000

The global optimum solution from LINGO indicates that the minimum compensatory power (FCOMP) is determined as 412.5 MW, in which coal plants 1, 3–5, as well as oil power plant 9 (indicated by binary variables of 1) are to be retrofitted with carbon capture technique. These resemble the results in Example 4.3 (see Table 4.5 for the summary of the results). A cross-check on CO_2 intensity reveals that its value has been reduced to 0.355 t/MWh (= 1,099.8 t/h/3,100 MW).

Further Reading

The simplified source–sink matching problem described in this chapter may not always reflect complications that arise in real systems. For example, temporal constraints may arise such that the operating lives of sources and sinks do not all coincide. More advanced *mixed integer linear program* (MILP) models that account for temporal constraints have been developed (e.g., Tapia et al., 2014; Shaik and Kumar, 2018). In this chapter, the problems of CO_2 balancing during source–sink matching and electricity balancing to account for the effects of retrofits are dealt with separately. The interested reader may explore the unification of these aspects into a single model as described by Lee et al. (2014). The MILP model in the latter work simultaneously accounts for CO_2 and electricity balancing. Yet another MILP model is also presented for Taiwan's energy sector for complex decision-making and cost trade-offs in the deployment of CC technologies and low-carbon energy sources; this is found in Chapter 10 of this book.

More recently, there has been significant interest in developing optimization models for integrated *carbon capture, utilization and storage* (CCUS) models with both CO_2 storage and utilization sinks. A survey of different mathematical models for planning CCS and CCUS systems can be found in a recent paper by Tapia et al. (2018).

References

Gau, C.-Y., Schrage, L. E. 2004. *Implementation and testing of a branch-and-bound based method for deterministic global optimization: operations research applications.* In Floudas, C. A., Pardalos, P. (Eds.), *Frontiers in Global Optimization.* Springer, Boston, MA, pp. 145–164.

Klemeš, J. J., Kravanja, Z. 2013. Forty years of heat integration: Pinch analysis (PA) and mathematical programming (MP). *Current Opinion in Chemical Engineering,* 2, 461–474.

Lee, J.-Y., Tan, R. R., Chen, C.-L. 2014. A unified model for the deployment of carbon capture and storage. *Applied Energy,* 121, 140–148.

Pękala, Ł. M., Tan, R. R., Foo, D. C. Y., Jeżowski, J. M. 2010. Optimal energy planning models with carbon footprint constraints. *Applied Energy,* 87(6), 1903–1910.

Shaik, M.A., Kumar, A. 2018. Simplified model for source-sink matching in carbon capture and storage systems. *Industrial and Engineering Chemistry,* 57, 3441–3442.

Tan, R. R., Foo, D. C. Y. 2007. Pinch analysis approach to carbon-constrained energy sector planning. *Energy,* 32(8), 1422–1429.

Tan, R. R., Ng, D. K. S., Foo, D. C. Y. 2009. Pinch analysis approach to carbon-constrained planning for sustainable power generation. *Journal of Cleaner Production,* 17(10), 940–944.

Tan, R. R., Ng, D. K. S., Foo, D. C. Y., Aviso, K. B. 2010. Crisp and Fuzzy integer programming models for optimal carbon sequestration retrofit in the power sector. *Chemical Engineering Research and Design,* 88(12), 1580–1588.

Tapia, J. F. D., Lee, J.-Y., Ooi, R. E. H., Foo, D. C. Y., Tan, R. R. 2014. Planning and scheduling of CO_2 capture, utilization and storage (CCUS) operations as a strip packing problem. *Process Safety and Environmental Protection,* 104, 358–372.

Tapia, J. F. D., Lee, J.-Y., Ooi, R. E. H., Foo, D. C. Y., Tan, R. R. 2018. A review of optimization and decision-making models for the planning of CO_2 capture, utilization and storage (CCUS) systems. *Sustainable Production and Consumption,* 13, 1–15.

Part B

Applications of Carbon Management Networks

6

Applications of Carbon Emission Pinch Analysis (CEPA) for China

Jia Xiaoping
Qingdao University of Science and Technology

Li Zhiwei
University of the Witwatersrand

Wang Fang
Qingdao University of Science and Technology

This chapter outlines how CEPA and its extensions serve as adequate tools for practical energy planning problems and municipal solid waste (MSW) management in China. Both applications of existing CEPA knowledge and newly developed ideas will be introduced.

6.1 Overview

For nearly three decades, Chinese society and the environment have been undergoing rapid and drastic changes. The term "ecological civilization" is becoming increasingly important in the global discourse, especially in China (Greene, 2017). These changes will have dramatic effects on economic development and its interactions with resources and environmental issues, such as energy, water, and emissions to air and water. Wang et al. (2019) projected that China's total emissions from fossil fuel and industrial processes will peak at 13–16 Gt/y CO_2 ahead of year 2030, based on the data from 50 Chinese cities. As the largest carbon emitter in the world, China's carbon management is of great significance to sustainable growth and global climate efforts.

Pinch analysis was originally developed as a systematic approach for optimizing energy use in the industry, by taking into account thermodynamic principles (Linnhoff et al., 1982). Tan and Foo (2007) first extended the use of pinch analysis into energy sector planning with carbon emission constraint. This method is now known as CEPA, and has been applied to many nations, including Ireland, New Zealand, the United States, and China. These applications are surveyed in a review paper (Foo and Tan, 2016). The first Chinese paper of CEPA was published by Jia et al. (2009), who performed carbon-constrained energy planning in a chemical industrial park, following the framework proposed by Tan and Foo (2007). From then on, researchers in China have played a key role in expanding the application of CEPA to different industrial sectors and at different scales. The reports of CEPA may be categorized as follows.

- Articles in Chinese journals

    According to a detailed literature survey, there are ten journal papers and two master's degree dissertations published in Chinese between 2009 and 2019. Most of these papers focused on the applications of CEPA to energy sector planning at different scales, i.e. building area, regional, provincial, and national levels. The list of selected cases where CEPA approaches have been used in different scales is shown in Table 6.1. All of these works were based on the method developed by Tan and Foo (2007) and later extended by Jia et al. (2009).

    The work of Zhang and Long (2011) and Zhu and Gong (2012) considers renewable energy sources in low-carbon community energy planning based on carbon emission reduction and limited resources. Many cases of provincial carbon constrained energy planning were reported. These provinces are Hubei (Hu, 2012), Tianjin (Yu et al. 2014), Jiangshu (Tang and Liao, 2014), Yunnan (Jiang 2014), and Sichuan (Xiang et al., 2016). At the national scale, Liang and Lu (2015) identified the energy bottleneck with CEPA and determined the minimum quantities of carbon-neutral energy in meeting the conditions of carbon emission constraints through energy substitution. Note that most authors in Chinese literature prefer to use the term "Carbon Pinch Analysis" than CEPA.

    Energy saving and emission reduction are the effective strategies to deal with the challenges of climate change and energy supply for a growing economy. These papers tried to support energy saving and carbon mitigation strategies at different scales. CEPA was used to determine the targets of carbon-neutral or low-carbon energy resources, based on the established graphical tools such as EPPD, or the equivalent algebraic tool.[1]

- Articles in international journals

    There are four papers published in international journals by Chinese researchers between 2009 and 2019, as shown in Table 6.2. The applications of CEPA were further extended from carbon-constrained energy planning to chemical industry (Qin et al., 2017), biomass supply chain (Li et al., 2016), and MSW management (Jia et al., 2018).

---

[1] See Chapter 2 for detailed procedure of EPPD and Chapter 3 for algebraic technique.

TABLE 6.1

CEPA-Related Work in China Published in Chinese Journals (2009–2019)

Year	Sector	Scale	Contributions	References
2009	Energy	Industrial park	• First CEPA paper in China; • Carbon constrained energy planning in a chemical industrial park	Jia et al. (2009)
2011	Energy	Community building area	• Renewable energy planning in building areas	Zhang and Long (2011)
2012	Energy	Regional	• Regional energy mix; Comparison with mathematical program	Yao and Wang (2012)
2012	Renewable	Community building area	• Renewable energy planning in building areas; • Considering marginal cost for energy utilization	Zhu and Gong (2012)
2012	Energy	Provincial	• Regional energy mix in Hubei	Hu (2012)
2013	Energy	Regional	• Regional energy mix	Zhao et al. (2013)
2014	Energy	Provincial	• Regional energy allocation in Tianjin; • Energy supply and carbon emission indices determination	Yu et al. (2014)
2014	Energy	Provincial	• Energy consumption structure optimization in Jiangsu	Tang and Liao (2014)
2014	Energy	Provincial	• Energy consumption structure optimization in Yunnan • Integration of CCS technology to reduce emissions	Jiang (2014)
2015	Energy	National	• Identification and solution of bottleneck energy uses; • Optimization of China's energy structure;	Liang and Lu (2015)
2016	Renewable	Provincial	• Renewable energy utilization in Sichuan	Xiang et al. (2016)
2017	Energy	Regional	• Regional energy allocation optimization	Qi et al. (2017)

Note that the promotion of renewable electricity sector at the regional level is still hampered by availability and reliability constraints. Li et al. (2016) developed a graphical pinch analysis-based approach for carbon-constrained electricity planning and supply chain synthesis of biomass energy at the regional level. Industrial structure adjustment and low carbon technology retrofit are the alternatives for policymakers and plants to simultaneously meet the energy demand and carbon emission limit. Qin et al. (2017) proposed

TABLE 6.2

CEPA-Related Work in China Published in International Journals (2009–2019)

Year	Sector	Scale	Contributions	References
2016	Renewable	Regional	• Carbon-constrained regional renewable electricity planning with Biomass Supply Pinch Analysis • A two-step approach to synthesis an optimal biomass supply chain network	Li et al. (2016)
2016	Electricity	National	• Multi-dimensional pinch analysis of the power sector • A more complete picture of the true environmental impact of China's power sector	Jia et al. (2016)
2017	Chemical industry	Sectoral	• Product-based Carbon Constraint Energy Planning combined with improved GM(1,1) model • Low-carbon development of methanol production industry.	Qin et al. (2017)
2018	Solid waste	Regional	• A hybrid LCCA and CEPA for MSW management • Different scenarios to generate optimized waste management results for 2020 and 2035 in Qingdao	Jia et al. (2018)

a product-based carbon-constrained energy planning for the methanol industry in 2020 in China. MSW system is one of the major greenhouse gas (GHG) emitters. The optimal planning for MSW system will facilitate to mitigate carbon emissions. Jia et al. (2018) developed an integrated method for planning a carbon-constrained MSW management system for Qingdao City.

In the following sections, two applications of CEPA in China are demonstrated. The first case considers how to use CEPA to minimize the emission of biomass supply chain at a regional scale. The second is to construct an MSW system with low-carbon emissions.

6.2 Biomass Supply Chain Network Synthesis

Renewable energy has an important role to play in Chinese power sector, especially at the regional level. China can potentially produce about 174.4–248.6 Mt/y of crop residues on a dry basis (Chen, 2016). It is estimated that rice straw accounts for about 47% of total residue production, while wheat straw accounts for about 25% (Chen, 2016). It is believed that once it becomes economically viable to produce crop residues, the supply of crop residues does not change much as biomass price increases. Corn stover and wheat straw can contribute 28% and 25%, respectively, to total biomass production in China (Chen, 2016). One promising option is to integrate renewable energies into the energy mix on a regional scale. A detailed case study of Laixi County in China is used to demonstrate the applicability of the CEPA-based approach for promoting renewable utilization at the regional level (Li et al., 2016).

6.2.1 Methodology

An extended graphical pinch analysis-based approach considers carbon-constrained regional electricity planning and supply chain synthesis of biomass energy. The minimum renewable energy target is first identified by CEPA. Next, a demand-driven approach is applied to synthesize a *biomass supply chain network* to meet the established target in a given region. The framework for synthesizing a regional biomass supply chain network is summarized as in Figure 6.1, with steps given as follows (Li et al., 2016).

Step 1: Identification of electricity target from renewable sources by CEPA.

An EPPD is used to determine the optimum allocation of power sources for a set of energy demands. Detailed steps for plotting the EPPD have been outlined in Chapter 2 of this book.

Step 2: Minimizing carbon footprint by Biomass Supply Pinch Analysis (BSPA).

The region is divided into many different geographic zones. Thus, the crop area in each zone can be estimated. Note that there is only one biomass

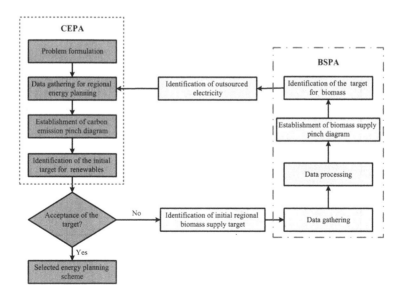

FIGURE 6.1
Procedure for synthesizing a regional biomass supply chain network (Li et al., 2016).

collection point in each zone. We then estimate the quantity of biomass source in each zone and determine the distance between different zones. Hence, the carbon footprint for the biomass transportation is identified. It is assumed that the carbon footprint of biomass allocation only exists as a result of transportation to different zones; they are assumed as zero if the biomass is utilized in the same zone where it is generated. The potential electricity for each zone (P_{elec}, GWh/y) and carbon footprint (CF) can be determined using Equations 6.1 and 6.2, respectively.

$$P_{elec} = M_{Bio} \times \lambda \times HLV \times \eta / 3,600 \tag{6.1}$$

$$CF = M_{Bio} \times Dist \times CEF_{fuel} \tag{6.2}$$

where HLV is the lower heating value for biomass, GJ/t; η is the thermal efficiency of power plant; CF is the carbon footprint, t CO_2-e/y; M_{bio} is the quantity of biomass, t/y; Dist is the distance of biomass collected point to the thermal plant, km; CEF_{fuel} is the carbon footprint emission factor, t CO_2-e/(t.km); λ is the collection efficiency. The latter is defined as the ratio of the actual biomass collected to its full potential in all zones. Note that in this work, only carbon emissions of transportation process is considered.

Biomass Supply Pinch Diagram (BSPD) is plotted with cumulative carbon emissions versus cumulative electricity, as shown in Figure 6.2. *Biomass*

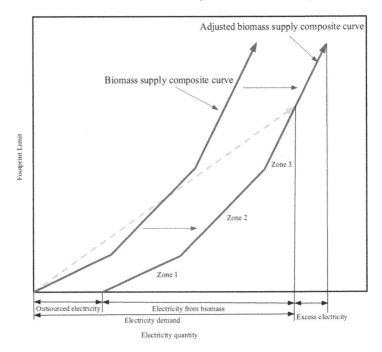

FIGURE 6.2
BSPD for synthesizing biomass supply chain network.

Supply Cumulative Curve (BSCC) is constructed on the BSPD. On the one hand, if the total amount of electricity from biomass is insufficient to satisfy the demand (dashed line in Figure 6.2), BSCC can be shifted horizontally to meet the demand. The opening on the left is the minimum outsourced electricity, which is to be supplied by coal-based electricity. Hence, it is necessary to redo CEPA with the updated results of biomass. On the other hand, the total amount of biomass electricity is sufficient to supply to the demand. The opening on the right corresponds to the excess energy, which cannot be utilized in view of emission limits. Therefore, the excess part should be deducted from the BSCC.

Case Study 1

Laixi County located in Shandong Province is selected as the case study. The county is divided into 15 zones. Data for biomass energy planning is given in Table 6.3 (Li et al. 2016). The electricity consumption for Laixi County is about 1,700 GWh in 2012. The electricity demand is expected to grow to 2,500 GWh by 2020. The geography data and potential biomass data for all 15 zones are shown in Table 6.4. In this case study, the collection efficiency (λ) is set as 50%, whilst the thermal efficiency (η) is fixed as 0.3.

In step 1, the target of electricity from biomass was determined as 750 GWh/y, as shown in the EPPD in Figure 6.3. Next, we will check if the amount of collected biomass is enough to meet the target.

The potential electricity for all the zones is shown in the column 2 of Table 6.4. Note that each zone has a different thermal value due to the different types of biomass. For instance, in zone 1, the total biomass for zone 1 is 39,136 t/y. The average thermal energy of the biomass is 15.23 GJ/t (Li et al., 2016), which is the production of thermal energy of specific biomass and its corresponding percent of the total biomass. With the biomass collection efficiency being set to 50%, the electricity from biomass is calculated using Equation 6.1 as 24.84 GWh/y (= 39,136 t/y $\times$ 0.5 $\times$ 15.23 GJ/t $\times$ 0.3/3,600). Therefore, the total electricity from the biomass for all zones is determined as 470 GWh/y (see last

TABLE 6.3

Data for Case Study

Energy Source	Emission Factor ($ktCO_2$-e/GWh)	Electricity for 2012 (GWh/y)	Electricity for 2020 (GWh/y)
Coal	1	1,430	1,070
Coal (with CCS)	0.2	0	500
Oil	0.8	180	0
Natural gas	0.5	0	180
Biomass	0	90	750
Total	--	1,700	2,500

TABLE 6.4

Coordinate for Zones in Laixi County

Z_i	Name	Location (km, km)	Biomass (t/y)	Low Heating Value (GJ/t)[a]	Z_i	Name	Location (km, km)	Biomass (t/y)	Low Heating Value (GJ/t)[a]
1	Wangcheng	(16, 16)	39,136	15.23	9	Rizhuang	(1, 30)	65,349	15.17
2	Shuiji	(16, 22)	29,007	15.28	10	Nanshu	(1, 40)	48,717	15.22
3	Jingkaiqu	(18, 28)	31,968	15.22	11	Hetoudian	(15, 38)	57,614	15.26
4	Guhe	(3, 21)	43,625	15.17	12	Dianpu	(0, 0)	33,357	15.28
5	Sunshou	(8, 10)	28,540	15.29	13	Liquanzhuang	(20, 0)	63,669	15.17
6	Jiangshan	(16, 2)	76,320	15.19	14	Wubei	(0, 20)	58,026	15.22
7	Xiagezhuang	(10,1)	65,147	15.14	15	Malianzhuang	(5, 42)	73,036	15.28
8	Yuanshangn	(0, 10)	33,885	15.16		Total	—	747,396	—

[a] Data compiled from Wei (2014).

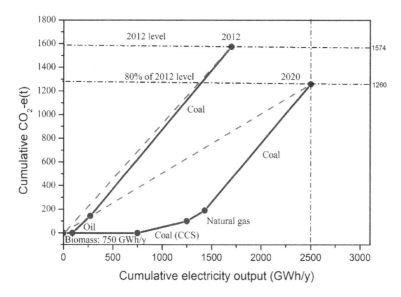

FIGURE 6.3

EPPD for Laixi County.

TABLE 6.5

Data for Scenario (Collection Efficiency of 50%)

Zone (i)	Potential Quantity (GWh/y)	Carbon Emission for Zone 2 (t CO_2-e/y)	Carbon Emission for Zone 13 (t CO_2-e/y)
1	24.84	5.69	15.63
2	18.47	0.00	15.71
3	20.27	4.89	21.72
4	27.57	13.77	28.53
5	18.18	9.96	10.79
6	48.30	36.96	8.26
7	41.09	34.45	15.85
8	21.41	16.41	18.35
9	41.30	26.89	56.18
10	30.90	27.63	52.23
11	36.63	22.03	53.46
12	21.23	21.97	16.15
13	40.24	34.47	0.00
14	36.79	22.65	39.74
15	46.49	40.36	78.86
Total	473.72	318.13	431.46

entry in Table 6.5; note that the value is rounded off for simplicity). Since the biomass sources come primarily as residues from farm, carbon emissions are assumed to be contributed by its transportation network. In this work, it is assumed that the biomass is transported from the collection points to the biomass energy conversion plants for electricity generation. The carbon emission factor (CEF_{fuel}) is set as 0.04 kg CO_2/t/km. We also made an assumption that power plants with a capacity of 100 MW are constructed in zone 2 or 13, which is due to the fact that the local government has the intention to build the thermal plant in these two zones. Hence, for both cases, biomasses from all other zones need to be sent to these zones for electricity generation. The distance between different zones is obtained based on the location data in Table 6.4. The corresponding carbon emissions for the transportation of biomass to zones 2 and 13 are determined using Equation 6.2.

Since the sum of electricity of biomass is 470 GWh/y, it is insufficient to meet the demand of 750 GW h/y, as being targeted by the EPPD (Figure 6.3).

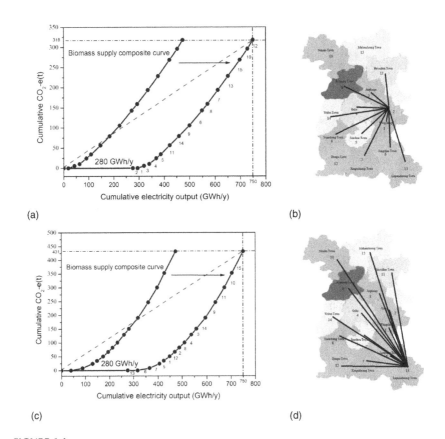

(a)   (b)

(c)   (d)

FIGURE 6.4
BSCC and biomass supply chain networks: (a) BSCC for zone 2, (b) biomass supply chain network for zone 2, (c) BSCC for zone 13, and (d) biomass supply chain network for zone 13.

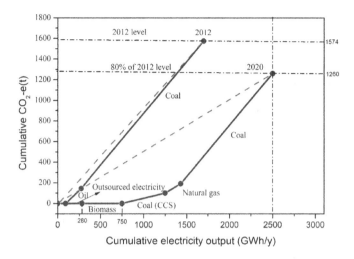

FIGURE 6.5
Final optimal energy mix profiles for Laixi County.

In order to meet the energy demand, the BSCC needs to be shifted horizontally, as shown in Figure 6.4a and c. The horizontal distance of BSCC to the origin is the target of outsourced energy with zero net emissions. Figure 6.4a and c shows the BSPD for both cases where biomasses are to be sent from all zones to power plants in zone 2 or 13. In order to meet the minimum carbon emission, all parts of the source curve should be chosen to meet the demand. As shown in Figure 6.4a and c, carbon emission of transportation for zones 2 and 13 are determined as 318 t CO_2-e/y and 431 t CO_2-e/y, respectively (given by the vertical distance of the BSCC). Note that there exists a significant gap of carbon emissions between these cases. Hence, from the viewpoint of carbon emission, the thermal plant is more appropriate to be located in zone 2, as it has lower carbon emission. The revised EPPD is shown in Figure 6.5, which shows that the electricity generated by biomass is determined as 470 GWh/y and the outsourced electricity as 280 GWh/y (=750–470 GWh/y).

6.3 MSW Management

MSW is defined as the most complicated solid waste streams resulting from the household/residential and the business/commercial sectors. With the continuous industrialization and urbanization, amounts of such wastes have tremendously increased over the last decades. Daily MSW generation per capita has increased from 448.3 g in 1980 to 653.2 g in 2014 in China (Gu et al., 2016). MSW systems are thus a significant source of GHG emissions, contributing about 5% of global GHG emissions in the form of CO_2, methane (CH_4), and nitrous oxide (N_2O). The disposal of MSW includes collection, transportation, sorting and recycling, landfill disposal, composting, and incineration. Thus, it can be seen that the reasonable planning of solid waste system is helpful to reduce GHG emissions in cities and achieve the reduction targets.

6.3.1 Methodology

A hybrid life cycle carbon emission accounting (LCCA)/CEPA framework for MSW management is developed, as shown in Figure 6.6. The system boundary of the research is defined to state the scope of the work. Next, the carbon emission accounting is presented to obtain the emission for the technologies of MSW treatment. Based on the development of technologies and authority preferences, MSW treatment scenarios are designed to meet the carbon emission target set by the local government.

- *MSW system boundary*
  The MSW management system boundary considers the life cycle of MSW. It includes the collection and transportation of waste, the sorting of recyclable waste, waste incineration, landfill, and other

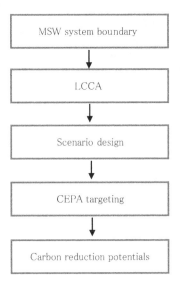

FIGURE 6.6
The framework for LCCA/CEPA.

stages of the system. Production and transportation of the various forms of energy consumed in MSW management can also produce GHGs, which are taken into account.

- *LCCA*

  The MSW system is simplified into the following stages: collection and transportation of waste, waste sorting and recycling, landfill, waste incineration, and energy use. The carbon emissions are considered from production and transportation of the various forms of energy consumed in this system. Material and energy consumption data for all stages of the life cycle and CEFs for each process are determined. Jia et al. (2018) presented the detailed carbon accounting model for each process. In this study, we convert all GHG emissions into CO_2 equivalents.

- *Scenario design*

  Seven different MSW technologies are considered, as shown in Table 6.6. The quantities of different types of solid wastes (food waste, metals, glass, paper, etc.) are allocated to the corresponding waste treatment technologies. Different mix of the treatment technologies results in different carbon emissions. Different scenarios for such mix were set to investigate optimal waste management option for 2020 and 2035. In these scenarios, the goals were meant to reduce carbon emissions by 20% in 2020 and by 45% in 2035. As new options become available, it will be important to increase the use of

TABLE 6.6

MSW Treatment Technologies and Their Descriptions

Tech.	Abbr.	Description
3R	A1	Recycling and reusing waste to generate new economic and social benefits.
Landfill (direct emissions)	A2	Landfill site operation with conventional equipment. Landfill *gas* (LFG) is released into the environment without any control.
Landfill (incineration)	A3	LFG is captured and sent into a flare system.
Landfill (electricity without CCS)	A4	LFG is captured and used as fuel to generate electricity.
Landfill (electricity with CCS)	A5	LFG is captured and used to generate electricity; CO_2 emissions from combustion are then captured for subsequent geological storage.
WtE (without CCS)	A6	Waste heat generated during incineration is supplied to the district.
WtE (with CCS)	A7	Waste heat generated during incineration is supplied to the district; CO_2 emissions from combustion are then captured for subsequent geological storage.

these low-carbon technologies. Such technological shifts will allow lower carbon emissions targets to be achieved.

- *CEPA targeting*

    After obtaining the carbon emissions of each technology, CEPA is used to optimize carbon emissions of the MSW system. The various treatment technologies result in a supply composite curve, which is plotted with the cumulative waste as the horizontal axis and the cumulative CO_2 as the vertical axis. The connection between the origin and the end point of supply composite curve is then seen as the demand composite curve. The pinch point refers to the intersection point of average demand composite curve and supply composite curve. Above the pinch point, the overhanging segment of supply composite curve is removed due to a higher emission factor.

- *Carbon reduction potential*

    Based on the results of CEPA, the amounts of carbon emissions produced by different treatment methods are analyzed under different scenarios. Such technological shifts will allow lower carbon emissions targets to be achieved. Feasible portfolios of different waste treatment technologies are recommended to construct the optimal MSW management strategy for the decision-making.

Case Study 2

The MSW system of Qingdao City is chosen as the case study. As of 2015, Qingdao had an urban population of 5,504,000. A total of 2.21 Mt of MSW was generated in 2015 (Kan, 2017). The per capita output of MSW in Qingdao City is about 1.1 kg/d. This amount will grow to 3.10 Mt in 2020, and 6.50 Mt in 2035 (Qu, 2011).

Due to space constraint, detailed steps of LCCA are not shown (please refer to Jia et al., 2018). The carbon emissions and waste for each technology are summarized in Table 6.7. The first step is to construct a set of supply composite curves according to the ascending CEFs. The demand composite curve is plotted based on the total quantity and emission target, which is a reduction of 20% carbon emission based on baseline, i.e., year 2015, as shown in Figure 6.7. If the carbon emission reduction target is to be achieved, the waste above the pinch point cannot be properly handled. The exceeding segment of supply composite curve needed to be replaced by the first slowest slope, which is A1. It will be replaced by the second slowest one if the proposal capacity exceeds the upper limit of A1, and so on, until all constraints are satisfied. Finally, the source composite curves that meet the emission reduction targets of years 2020 and 2035 are obtained.

Under the reduction targets, the carbon emissions in 2020 come to 1.53 Mt/y. The major contributor to carbon emissions in all three scenarios in 2020 is A2, which accounts for 0.85 Mt/y. Another main contributor of CO_2

TABLE 6.7

Carbon Emissions vs. Waste for Each Technology for 2020 and 2035

	Year 2020		Year 2035	
Tech.	Waste x_1 (t/y)	Carbon Emissions y_1 (t/y)	Waste x_2 (t/y)	Carbon Emissions y_2 (t/y)
A1	147,000	0	886,000	0
A5	39,000	2,000	284,000	13,000
A7	93,000	12,000	1,235,000	272,000
A4	837,000	109,000	2,080,000	441,000
A3	403,000	64,000	1,495,000	236,000
A6	682,000	494,000	520,000	377,000
A2	899,000	852,000	-	-

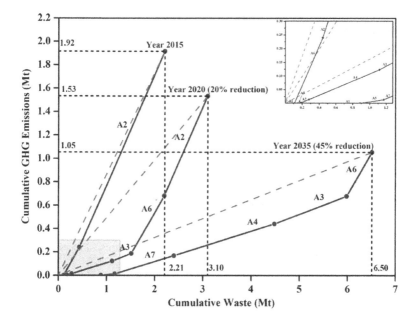

FIGURE 6.7

Source composite curves to meet emission reduction targets of 2015, 2020, and 2035.

is A6, which produced 0.49 Mt/y. With a 45% reduction in carbon emissions since 2015, carbon emissions in 2035 are determined as 1.05 Mt/y. The contributions to carbon emissions ranked from highest to lowest in scenario as follows: A6, A4, A3, A7, A5, and A1. Among them, A5 and A1 are shown in the small block diagram in Figure 6.7. Because of the high intensive carbon emissions of A6, the processing capacity of A6 must be reduced to achieve lower carbon emissions in the future.

6.4 Conclusion

CEPA method has made a good contribution to the energy planning and related issues in specific Chinese provinces, cities, and industries. An important feature of this graphical method is its capability of showing useful insights of the system under study. With regard to Chinese carbon emission mitigation, much more research effort could be expanded to the national energy planning considering the Paris Agreement carbon emission control constraint. Moreover, a further application of this method which may be explored is to examine the interrelation between energy development and industry transfer in detail.

Acknowledgment

The authors would like to thank the financial support provided by the National Natural Science Foundation of China (no. 41771575).

References

Chen, X., 2016. Economic potential of biomass supply from crop residues in China. *Applied Energy*, 166, 141–149.

Foo, D. C. Y., Tan, R. R. 2016. A review on process integration techniques for carbon emissions and environmental footprint problems. *Process Safety and Environmental Protection*, 103, 291–307.

Greene, H. 2017. Ecological civilization and the 19th National Congress of the Communist Party of China. Available at: www.ecozoicstudies.org/wp-content/uploads/2017/10/Ecological-Civilization-and-19th-Congress-China.Greene-ER-2017-0910.pdf.

Gu, B., Jiang, S., Wang, H., Wang, Z., Jia, R., Yang, J., He, S., Cheng, R. 2016. Characterization, quantification and management of China's municipal solid waste in spatiotemporal distributions: A review. *Waste Manage*, 61, 67–77.

Hu, X. 2012. Energy consumption structure optimization of Hubei province based on carbon emission pinch analysis. Master Degree Thesis. Huazhong University of Science and Technology, Wuhan, China (in Chinese).

Jia, X., Liu, C., Qian, Y. 2009. Carbon emission pinch analysis for energy planning in chemical industrial park. *Xiandai Huagong/Modern Chemical Industry*, 29(9), 81–85 (in Chinese).

Jia, X., Li, Z., Wang, F., Foo, D. C. Y., Tan, R. R., 2016. Multi-dimensional pinch analysis for sustainable power generation sector planning in China. *Journal of Cleaner Production*, 112, 2756–2771.

Jia, X., Wang, S., Li, Z., Wang, F., Tan, R.R., Qian, Y. 2018. Pinch analysis of GHG mitigation strategies for municipal solid waste management: A case study on Qingdao City. *Journal of Cleaner Production*, 174, 933–944.

Jiang, J. 2014. Energy structure optimization of Yunnan province based on carbon emission pinch analysis. Master Degree Thesis, Yunan University of Finance and Economics, Kunming, China (in Chinese).

Kan, B., 2017. Study on present situation, physicochemical characteristics and disposal methods of municipal solid waste in Qingdao City--Data Analysis from 2004 to 2016. Master Degree Thesis, Qingdao University, Qingdao, China (in Chinese).

Li, Z., Jia, X., Foo, D. C. Y., Tan, R. R. 2016. Minimizing carbon footprint using pinch analysis: The case of regional renewable electricity planning in China. *Applied Energy*, 184, 1051–1062.

Liang, L., Lu, Q. 2015. Optimization of China's energy structure based on carbon pinch analysis. *Resources Science*, 37(2), 291–298 (in Chinese).

Linnhoff, B., Townsend, D. W., Boland, D., Hewitt, F., Thomas, A., Guy, A. R., Marsland, P. H. 1982. *User Guide on Process Integration for the Efficient Use of Energy*. Institution of Chemical Engineers, Rugby.

Qi, C., Li, T., Liu, J., Zhang, L., 2017. Optimization of carbon pinch technology for community energy planning. *Building Energy Efficiency*, 45(12), 59–63 (in Chinese).

Qin, Z., Tang, K., Wu, X., Yu, Y., Zhang, Z., 2017. Product-based Carbon Constraint Energy Planning with pinch analysis for sustainable methanol industry in China. *Chemical Engineering Transactions*, 61, 103–108.

Qu, X. Z. 2011. Life cycle assessment of urban solid waste processing systems in Qingdao. Master Degree Thesis. Qingdao Technological University, Qingdao, China (in Chinese).

Tan, R. R., Foo, D. C. Y. 2007. Pinch analysis approach to carbon-constrained energy sector planning. *Energy*, 32(8), 1422–1429.

Tang, J., Liao, X., 2014. Optimization of energy consumption structure in Jiangsu based on carbon pinch technology. *Resources Science*, 36(12), 2560–2568 (in Chinese).

Wang, H., Lu, X., Deng, Y., Sun, Y., Nielsen, C. P., Liu, Y., Zhu, G., Bu, M., Bi, J., McElroy, M. B., 2019. China's CO2 peak before 2030 implied from characteristics and growth of cities. *Nature Sustainability*, 2, 748–754

Wei, Q. Y.. 2014. Research on supply chain logistics cost of straw for biomass power generation. Ph.D. thesis. China Agricultural University, Beijing, China, 2014 (in Chinese).

Xiang, B., Yuan, Y., Yang, X., Cao, X., Yu, N., 2016. Renewable energy utilization for Sichuan province based on carbon pinch analysis. *Refrigeration & Air Conditioning*, 30(5), 615–621 (in Chinese).

Yao, M., Wang, C., 2012. Pinch analysis for regional energy allocation with carbon emission constraints. *Journal of Shanghai Maritime University*, 33(3), 58–63 (in Chinese).

Yu, H., Wang, J., Hu, L., 2014. Study on regional energy allocation under carbon emission constrained in Tianjin: based on analysis of carbon pinch. *Environmental Pollution & Control*, 36(6), 90–95 (in Chinese).

Zhang, G., Long, W. 2011. Study of carbon pinch technology in community energy planning. In *Proceeding of 7th International Conference for Green Building and Energy Conservation, Beijing*, pp. 450–456 (in Chinese).

Zhao, L., Zhang, L., Cang, D., Li, Y., Liu, X. 2013. Application of Carbon pinch tech-
    nology in energy conservation. *Energy for Metallurgical Industry*, 32(2), 3–6
    (in Chinese).
Zhu, Y., Gong, Y. 2012. Carbon pinch analysis considering the economics of renew-
    able energy utilization. *Building Energy & Environment*, 31(3), 31–34 (in Chinese).

7

Applications of Carbon Emission Pinch Analysis (CEPA) for Indian Electricity Sector

Sheetal Jain and Santanu Bandyopadhyay

Indian Institute of Technology Bombay

7.1 Introduction

India, being the second most populous country (United Nations Census Bureau, 2019) and the fastest-growing economy in the world (Chazen Global Insights, 2016), requires a large amount of energy (on an average, 1,137 billion kWh of electric energy per year (WorldData.info)) for its sustenance. Due to the ever-increasing energy demand, with compounded annual growth rate (CAGR) of 7.82% during 2007–2008 to 2016–2017 (Central Statistics Office, 2018), proper planning of electricity sector is essential for the overall sustainable development. The electricity sector of India is dominated by fossil fuels, particularly coal from the 19th century till date, and gearing towards renewable energy sources (RES) in recent years. The share of renewable in India in 2001 was merely about 2.1% of the total installed capacity whereas, in year 2018, it rose to 21.2% (Ministry of Power, 2019a), a phenomenal increase. The fossil fuel-driven economy has many repercussions for the country in particular and the planet in general. In 2004, the per capita CO_2 emissions in India were reported to be 1.022 t/y and by 2014, it increased to 1.728 t/y (The World Bank, 2019) owning to India's growing economy and improved standards of living. Along with the emissions of CO_2, a major concern behind global warming and climate change, fossil fuel-based electricity generation also emit other harmful greenhouse gases (GHGs) like methane (CH_4) and nitrous oxide (N_2O). These gases affect the environment in many ways, by degrading the air quality, destroying the historical monuments and sites, causing acid rain, etc.

Apart from problems related to emissions, the cost-effectiveness of electricity generation, transportation, and distribution is also a significant concern for the nation. The lack of adequate infrastructure and inefficient distribution of electricity leads to energy losses, which in turn makes the nation electricity deficient. Different policies and frameworks were formulated by the Government of India (GoI) to provide sustainable and reliable power to the country. As can be seen later in this chapter that to achieve the sustainability in a cost-efficient manner, it is necessary to determine an optimal mix of energy from different fuel sources, in order to minimize emissions while taking into account various sustainability indicators, e.g., land footprint, water footprint, risk to humans, etc. The accountability of these sustainability indicators is very crucial in the planning of policies related to climate change adaptation and mitigation. It is particularly important in the developing countries like India where flash floods, seawater encroachments, droughts, etc. are the major concerns with the escalating effects of climate change. As introduced earlier in this book, *Carbon Emissions Pinch Analysis* (CEPA) is one of the efficient techniques to optimize the energy mix to improve the overall sustainability of the sector cost-effectively.

This chapter provides an overview of the Indian electricity sector and emphasis upon the local issues and problems faced by the sector due to inadequate infrastructure, unavailability of clean fuel for electricity generation, and losses in the distribution system. The contributions and importance of CEPA in the context of the Indian electricity sector are also discussed.

7.2 History of the Indian Electricity Sector

Electricity was first introduced in India in 1879 in Calcutta, the erstwhile capital of British India and presently known as Kolkata (Das, 2009). Afterward in 1880s, it reached the streets of Bombay (in the present day, it is called Mumbai), the financial capital of present India (Kale, 2014). Initially, electricity was supplied to only a few urban centers, offices, and ports while the rural areas and villages remain un-electrified. The primary source of electricity generation at that time was thermal and hydropower plants. Later in the 1960s, nuclear power plants were also introduced; however, the share of nuclear energy in electricity generation remained minimal. The government recognized the impact of the electricity sector on the nation's economy, and hence after independence (year 1947), improvements in the electricity sector became the main attention for the GoI. The installed capacity increased from 1.7 GW in 1950 to 89 GW in 1998, while the electricity generation increased from 5.1 to 420 billion kWh/y, and per capita consumption of electricity increased from 15 to 338 kWh/y for the same period (Indian Power Sector.Com, 2014). Despite the tremendous progress in the electricity sector, bridging the gap between the supply and demand remained a significant challenge for the GoI, as can be seen in Table 7.1

TABLE 7.1

Year-Wise Power Supply Shortages in India between 1992 and 2018 (Ministry of Power, 1998, 2019b)

	Energy (Billion Unit Net)				Peak (GW)			
Year	Requirement	Availability	Shortage	(%)	Peak Demand	Peak Demand Met	Shortage	(%)
1992–1993	305.3	279.8	25.5	8.4	52.8	42.0	10.8	20.5
1995–1996	389.7	354.1	35.6	9.1	61.0	49.8	11.2	18.4
2010–2011	861.6	788.4	73.2	8.5	122.3	110.3	12.0	9.8
2015–2016	1,114.4	1,090.9	23.5	2.1	153.4	148.5	4.9	3.2
2018–2019	1,274.6	1,267.5	7.1	0.6	177.0	175.5	1.5	0.8

(Ministry of Power, 1998). Other main challenges that the Indian electricity sector faced at that time were related to transmission and distribution (T&D) losses, and electrification of rural areas and over 590,000 villages (Ministry of Power, 1998).

During that time, most commissions considered the issues caused by the generation of electricity from different power plants (dependence on fossil fuels, pollution generation, emissions of harmful gases, environmental impacts, etc.) as issues "to be tackled later" (Dubash and Rajan, 2001). These issues were not the primary concerns and they came into light only after the adoption of international environmental treaties like the United Nations Framework Convention on Climate Change (UNFCCC), Kyoto Protocol, and the Paris agreement (COP21). These treaties and agreements aim to "stabilize greenhouse gas concentrations in the atmosphere at a level that would prevent dangerous anthropogenic interference with the climate system" (United Nations, 1992). The GoI then realized the importance of efficient use of energy in different sectors in order to reduce the environmental impacts caused by emissions of GHGs in generation, transmission, and utilization of energy. In September 2001, the Energy Conservation Act, 2001 (Ministry of Environment & Forests, 2004), was enacted by the GoI, which aimed to promote energy efficiency and conservation in all sectors. One of the features of this act was to renovate and modernize the old thermal power stations to increase their energy efficiency and hence leads to reduced GHG emissions. To offset the need for new power plants, focus is also directed to reduce the T&D losses. Conferences such as "Coal and Electricity in India" and "Thermal Power Generation – Best Practices and Future Technologies" (Ministry of

Environment & Forests, 2004) served as platforms to discuss about the present and the future of sustainability in Indian power sector. India presented its Initial National Communication to the UNFCCC in 2004 (Ministry of Environment & Forests, 2004) which aimed to reduce the GHG concentration in different sectors of the society. In COP21, the GoI committed to (International Energy Agency, 2015):

- Reduce the emissions intensity of its GDP by 33%–35% from 2005 levels by 2030.
- Achieve about 40% cumulative electric power installed capacity from non-fossil fuel-based energy resources by year 2030, with the help of the technology transfer and low-cost international finance including from Green Climate Fund (GCF).
- Create an additional carbon sink of 2.5–3 Gt of CO_2 equivalent through additional forest and tree cover by 2030.
- Better adaptation to climate change by enhancing investments in development programs in sectors vulnerable to climate change, particularly agriculture, water resources, Himalayan region, coastal regions, health, and disaster management.

7.3 Renewable Energy in India

Along with energy efficiency, the transition from conventional fossil fuels to RES was seen as one of the pathways for sustainable development of the electricity sector. It was estimated that India has a renewable energy potential of about 900 GW from commercially exploitable sources, i.e., 102 GW wind power (at 80-m mast height), 20 GW of small hydro, 25 GW of bio-energy, and 750 GW of solar power (Ministry of New and Renewable Energy, 2017). In the Electricity Act 2003, the focus is directed towards the promotion of new and RES in the power sector (Section 61) and towards developing a policy framework for stand-alone systems based on renewable and other non-conventional sources of energy for rural areas (Section 4) (Ministry of Law and Justice, 2003). These policies and regulations increased the share of renewables in India from 1.06% in 1999 (Ministry of Power, 2000) to 9.84% in 2009 (Ministry of Power, 2010) and further to 18.8% in 2017 (Ministry of Power, 2018). The year-wise distribution of installed capacity of different energy sources is given in Figure 7.1 (Ministry of Power, 2019a). Wind energy was the only non-conventional source of energy which played a role in electricity generation until 2004. However, after 2004, other RESs (excluding large hydropower plants) became more diverse by incorporating small hydroprojects, biomass gasifiers, biomass power, urban and

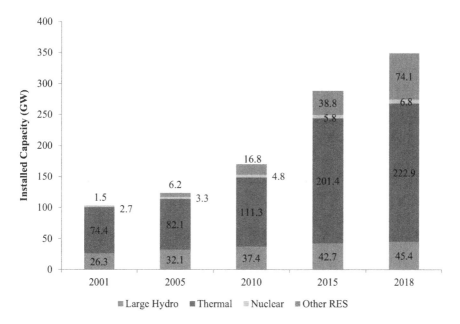

FIGURE 7.1

Installed capacity (in GW) of power stations in India (Ministry of Power, 2019a,b).

industrial waste power, and wind energy for power generation. After 2006, solar energy also became a part of electricity generation in India. Despite the tremendous increase in installed capacity (233.75%) from 2001 to 2018 (calculated from Figure 7.1), India is still an electricity-deficit country as can be seen in Table 7.1.

7.4 Policies and Schemes

The GoI has announced many policies to combat the challenges of T&D losses, rural electrification, and to ensure the availability of reliable and quality power to the nation. Some of the relevant schemes launched by the GoI are listed in Table 7.2. The Honorable Prime Minister Shri Narendra Modi launched Unnat Jyoti by Affordable LEDs and Appliances for All (UJALA) scheme by saying "I urge you all to use L.E.D. bulbs, save money, save energy and take part in helping our nation" (Energy Efficiency Services Limited, 2019). Apart from the schemes mentioned in Table 7.2, the GoI also launched an application called "TARANG" (Transmission App for Real-Time Monitoring & Growth) for real-time monitoring of transmission systems.

TABLE 7.2

Policies and Schemes Launched by the GoI for the Electricity Sector

S. No.	Name	Year	Aim
1	Deendayal Upadhyaya Gram Jyoti Yojana (DDUGJY) (Ministry of Power, 2014)	December 2014	• Electrification of 1,21,225 un-electrified villages • Intensive electrification of 5,92,979 partially electrified villages • Free electricity connections to 39.75 million Below Poverty Line (BPL) rural households
2	Pradhan Mantri Sahaj Bijli Har Ghar Yojana (Saubhagya) (Ministry of Power, 2017)	October 2017	• To achieve universal household electrification by providing last-mile connectivity • Electricity connection to all households in rural and urban areas
3	Ujwal DISCOM Assurance Yojana (UDAY) (Ministry of Power, 2015)	November 2015	• To track and reduce the T&D losses • To reduce the power losses by the installation of smart meters and allowing remote reading • Encourages public participation for the reduction of losses
4	Unnat Jyoti by Affordable LEDs and Appliances for All (UJALA) (Energy Efficiency Services Limited, 2019)	2015	• To promote efficient lighting • Enhance awareness on using efficient equipment which reduces electricity bills and helps preserve the environment

7.5 Carbon Emissions Pinch Analysis

The policies and frameworks discussed in the previous section help the nation for developing a future vision for more sustainable, economical, and environmentally-friendly energy generation and distribution. India's electricity sector is already very diverse and dramatic both in terms of scale and fuel mix. Moreover, various governmental initiatives like "Make in India" put extra pressure on making the electricity sector more efficient, reliable, and diverse in terms of technology, fuel mix, and scale (Buckley and Shah, 2017).

7.5.1 Single Objective CEPA

Apart from governmental efforts, academia is also concerned with the increasing level of emissions due to energy generation. Moreover, for a developing country like India, capital investment on commissioning of new power plants that are equipped with advanced technologies for emission reduction poses a significant challenge. Therefore, it is necessary to optimize the energy mix from existing power plants to meet the electricity demand of the nation and to maintain the emissions below a specified limit.

For India, the electricity demand is predicted to be 2,940 TWh for 2030 (Khosla and Dubash, 2015), and the emission target is assumed to be 1,420 Mt of CO_2 which is equivalent to 0.48 Mt/TWh of CO_2 (=1,420 Mt/2,940 TWh) (Krishna Priya, 2016). The energy mix from various existing power plants (Table 7.3) is optimized to meet the future energy demand of India, illustrated in Table 7.4, using the algebraic targeting technique (Foo et al., 2008).[1] Solar photovoltaic (PV) (CO_2 emission of 0.15 Mt/TWh and capital cost of 0.84 million $/MW) is considered as the available resource for future power plants (Krishna Priya, 2016). It is observed from Table 7.4 that the energy requirement from the future solar PV power plant is 1,298.51 TWh and the capital cost required is $622.57 billion.

In practice, there may exist many other sources of power generation apart from solar PV for satisfying the future energy demands (Table 7.5). The energy mix can be optimized using existing and the future power plants with the concept of prioritized cost (PC_k; Equation 7.1) developed by Shenoy and Bandyopadhyay (2007) for problems involving multiple resources.

$$\text{Prioritized cost}(PC_k) = \frac{cc_{rk}}{(C_{PN} - C_{rk})} \tag{7.1}$$

where cc_{rk} and C_{rk} represent capital cost and quality (CO_2 emission) associated with kth power plant and C_{PN} represents the Pinch quality. Prioritized cost captures the trade-off between capital cost and quality of the resource. It is proved by Shenoy and Bandyopadhyay (2007) that the introduction of kth resource in the energy mix is beneficial for capital cost minimization

TABLE 7.3

Data for Existing Power Plants

Source	Limit (MW) (Ministry of Power, 2019c)	CO_2 Emissions (Mt/TWh) (Krishna Priya, 2016)
Coal	195,810	1.08
Gas	24,940	0.45
Diesel	640	0.65
Nuclear	6,780	0
Hydro	45,399	0
Wind	36,690	0
Solar	30,070	0
Small Hydropower (SHP)	4,610	0
Biomass	9,270	0.11

---

[1] See Chapter 3 for detailed steps.

TABLE 7.4

Determination of Capacity Required for Future Power Plants (Solar PV) in India

C_k (Mt/TWh)	ΔC_k (Mt/TWh)	$\Sigma_j F_{D,j}$ (TWh)	$\Sigma_i F_{S,i}$ (TWh)	$F_{Net,k}$ (TWh)	Cum. $F_{Net,k}$ (TWh)	ΔE_k (Mt)	Cum. ΔE_k (Mt)
0			360.15	360.15			
	0.11				360.15	39.62	
0.11			56.84	56.84			39.62
	0.04				416.99	16.68	
0.15			1,298.51	1,298.51			56.3
	0.3		(F_{RE})		1,715.5	514.65	
0.45			174.78	174.78			570.95
	0.03				1,890	56.71	
0.48		2,940		−2,940			627.66
	0.17				−1,049.71	−178.45	
0.65			5.05	5.05			449.21
	0.43				−1,044.67	−449.21	
1.08			1,372.24	1,372.24			0 (C_{PN})
	8.92				327.57	2,921.92	
10				(F_{EX})			2,921.92

TABLE 7.5

Future Power Sources for Indian Power Sector (Krishna Priya, 2016)

	Limit (MW)	CO_2 Emissions (Mt/TWh)	Water Footprint (m³/MWh)	Cost (10⁶$/MW)
Coal	NA	1.08	2.56	0.56
Nuclear	9,550	0.02	2.80	0.73
Wind	47,000	0.07	0.015	1.35
Solar PV	NA	0.15	0.004	0.84
Solar Thermal	NA	0.185	1.08	1.68
Hydro	148,700	0.12	5.30	0.91
Small Hydro	15,000	0.12	5.30	1.73
Biomass	19,500	0.11	1.80	0.42
Coal with CCS	NA	0.10	5.03	0.73

only if its prioritized cost is less than that of all the better quality resources. The prioritized costs for all future energy sources are calculated using Equation 7.1 and tabulated in Table 7.6, calculated using Pinch quality (C_{PN}) of 1.08 Mt/TWh (obtained through the best quality resource in Table 7.4). As observed from Table 7.6, biomass has the least prioritized cost, followed by nuclear, coal with CCS, hydro, small hydro, solar PV, solar thermal, wind. It may be noted from Table 7.5 that both biomass and nuclear have limited capacity to add. On the other hand, coal with CCS does not have any capacity

TABLE 7.6

Prioritized Cost for Future Power Plants

Source	Prioritized Cost, PC_k (10^6 $/Mt)
Coal	-
Nuclear	87.35
Wind	1089.88
Solar PV	515.54
Solar Thermal	1,071.4
Hydro	216.42
Small Hydro	411.43
Biomass	70.61
Coal with CCS	94.48

limitation. Following the previous procedure, 119.57 TWh of biomass (F_{bio}), 75.29 TWh of nuclear (F_{Nuc}), and 1,023.49 TWh of coal with CCS (F_{CL-CCS}) are added to completely satisfy the energy demands. Targeting for each resource can be done individually, one at a time in succeeding order of prioritized cost. Due to brevity, only the final targeting results are tabulated in Table 7.7. The total capital cost required for commissioning of these new power plants is determined as $109.93 billion ($8.19 billion for biomass, $6.97 billion for

TABLE 7.7

Determination of Capacity Required for Different Power Plants

C_k (Mt/TWh)	ΔC_k (Mt/TWh)	$\Sigma_j F_{D,j}$ (TWh)	$\Sigma_i F_{S,i}$ (TWh)	$F_{Net,k}$ (TWh)	Cum. $F_{Net,k}$ (TWh)	ΔE_k (Mt)	Cum. ΔE_k (Mt)
0			360.15	360.15			
	0.02				360.15	7.2	
0.02			75.29	75.29			7.2
	0.08		(F_{Nuc})		435.44	34.84	
0.1			1,023.49	1,023.49			42.04
	0.01		(F_{CL-CCS})		1,458.93	14.59	
0.11			56.84 + [119.57 (F_{Bio})]	176.42			56.63
	0.34				1,635.35	556.02	
0.45			174.78	174.78			612.65
	0.03				1,810.13	59.72	
0.48		2,940		−2940			672.37
	0.17				−1,129.87	−188.7	
0.65			5.05	5.05			483.67
	0.43				−1,124.82	−483.67	
1.08			1,372.24	1,372.24			0
	8.92				247.42	2206.93	(C_{PN})
10					(F_{EX})		2203.93

nuclear, and \$94.77 billion for coal with CCS). By introducing different types of new power plants capital cost investment can be reduced by 82.34% (=(622.57–109.93)/109.93, as compared to the case when only solar PV is used.

7.5.2 Multi-Objective CEPA

Apart from cost-effectiveness, many other sustainability indicators such as land footprint, water footprint, energy return on investment (EROI), and risk to humans (see details in Table 7.8) play significant roles in determining the sustainability of energy generation. Therefore, it is crucial to determine the optimal energy mix from different power plants that meet the energy needs of the nation, while being economical and sustainable. Many researchers worked in this area in order to determine the optimal energy mix. For example Chandrayan and Bandyopadhyay (2014) optimized the cost of energy mix for transportation and industrial sectors of India, taking into consideration the CO_2 emissions limit with segregated targeting problems. Patole et al. (2017) used multiple index Pinch Analysis approach to optimize the energy mix considering multiple sustainability indicators, and combined them into a single sustainability indicator using Analytical Hierarchy Process (AHP); the latter uses expert's opinion in determining the relative importance of one sustainability criterion over the other. The authors observed that for the Indian power sector, CO_2 emissions and EROI

TABLE 7.8

Different Sustainability Indicators and Their Policy Relevance

Sustainability Index	Description (Patole et al. 2017)	Policy Relevance
Carbon footprint	Measures the contribution of an energy system to climate change	• To reduce and maintain CO_2 emissions for climate change adaptation and mitigation
EROI	Measures the energy productivity of a system	• To enhance the productivity of the sector • To maintain the economics of the sector
Land footprint	Measures the geographic area occupied by an energy system	• To manage social issues about land acquisition, rehabilitation, and resettlement
Water footprint	Measures the water stress caused by an energy system	• To avoid water stress in the area where the power plant is located • To lessen the intensity of freshwater use
Risk to humans	A qualitative measure of potential harm to inhabitants through pollutants, accidents, etc.	• To avoid the untimely deaths caused due to emissions • To avoid accidents at the power plant site

influence the energy mix more than land footprint, water footprint, or risk to humans. Krishna Priya and Bandyopadhyay (2017) optimized energy mix using Pinch Analysis approach by considering multi-objectives like water footprint and capital investment and at the same time maintaining the CO_2 emissions below the pre-defined limit.

Krishna Priya and Bandyopadhyay (2017) considered the existing power plants and the resources available for future power generation for satisfying the energy demand. Since it is a multiple objective optimization problem, the weighted sum method is adopted to combine multiple objectives into a single objective function, and then Pinch Analysis approach is applied to generate the Pareto-optimal front for the problem. Additionally, as multiple resources are considered to meet the demand, the concept of prioritized cost developed by Shenoy and Bandyopadhyay (2007) is adopted, and a quantity called *multiple objective prioritized cost* (MOPC) (Equation 7.2) is derived to govern the prioritization of the commissioning of new power plants.

$$\text{MOPC} = \frac{wC_{\phi_k} + (1-w)C_{\Psi_k}}{(C_{PN} - C_{rk})} \tag{7.2}$$

where w is the weighing factor, ϕ_k and Ψ_k are the two objective functions of the kth resource, C_{ϕ_k} and C_{Ψ_k} are the cost coefficients associated with these objective functions, C_{PN} is the Pinch quality (obtained using the best quality resource), and C_{rk} is the quality associated with the kth resource. Note that data for this study is adopted from Krishna Priya and Bandyopadhyay (2017). For this case, the objective functions that were taken into consideration are capital cost and water footprint minimization. Krishna Priya and Bandyopadhyay (2017) calculated the MOPCs for each resource and plotted the MOPC vs. weighing factor graph to determine the prioritizing sequence of the resources (Figure 7.2). The point of intersection of these lines identifies the weighing factor at which the prioritizing sequence may change. Though not depicted in Figure 7.2, up to a very low weighing factor $w = 0.001$, solar energy has the lowest MOPC, so the solution includes only solar power plants. After that, until point A ($w = 0.116$), the prioritizing sequence is [wind-solar]. Then the sequence changes to [wind-solar-biomass] from A to B ($0.116 \le w \le 0.154$) and to [wind-solar-nuclear-biomass] from B to C ($0.154 \le w \le 0.263$). Similarly, all the prioritizing sequences are obtained, and the Pareto-optimal front is generated (Figure 7.3).

The Pareto-optimal front in Figure 7.3 consists of six unique points. As observed from Figure 7.3, if the objective is only to minimize capital cost, coal with carbon capture and storage (CCS) and biomass plays an important role, however, as the objective shifts towards water footprint minimization, wind, and solar energy started playing their role. When the objective is more focused on water footprint minimization, coal with CCS is eliminated

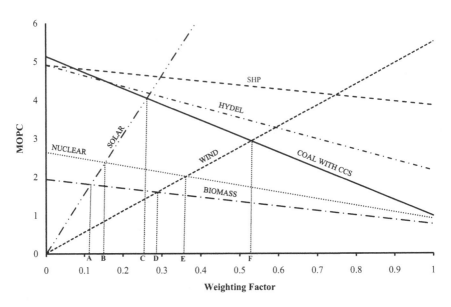

FIGURE 7.2
Prioritized versus weighing factor graph for the resources (Krishna Priya and Bandyopadhyay, 2017).

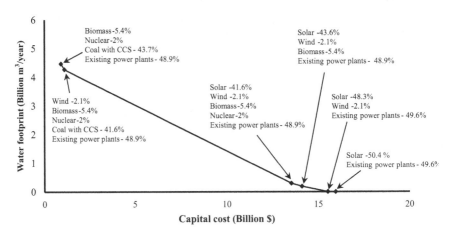

FIGURE 7.3
Pareto-optimal front for the Indian power sector (Krishna Priya and Bandyopadhyay, 2017).

from the energy mix, and solar energy takes over. It can be concluded from Figure 7.3 that coal with CCS and solar energy are the major contributors in the energy mix. Improvements in the cost-effectiveness and water foot-print reduction of these two major technologies can prove to be of significant importance for the Indian power sector.

These research works can serve as a base for the GoI for decision-making information and implementation of new policies and frameworks, for the sustainability of Indian power sector. Considering the complex and diverse nature of Indian power sector, there is a need for more realistic, complex, and concrete research focused on sustainability of Indian power sector which combines the expertise of academicians and the implementation knowledge of the government officials.

7.6 Conclusions

The fossil fuel dominated electricity sector of a developing country like India poses a serious threat towards global sustainability in terms of increasing CO_2 emissions and hence contributing more towards global warming. The GoI realized the repercussions of the fossil fuel-based electricity sector and implemented various policies and framework to reduce the emissions caused by the fossil fuel by emphasizing on energy efficiency and introducing renewable sources of energy in the sector. The fast-paced growth of renewable energy along with the modifications in the current fossil fuel-based power plants helps in providing affordable, reliable, and sustainable power to the nation. It is concluded that along with affordability and emissions reduction, there exist many factors responsible for the sustainability of the power sector like land footprint, water footprint, EROI, risk to humans, etc. It is also observed that the evolving economy needs a much higher share of renewable energy and shutdown of existing coal power plants for its sustainable existence. Such scientific research-based observations along with the hands-on experience and knowledge of government officials help in decision making for the policymakers and hence contribute towards sustainability.

References

Buckley, T., Shah, K. 2017. India's electricity sector transformation. Institute for Energy Economics and Financial Analysis. http://ieefa.org/wp-content/uploads/2017/11/India-Electricity-Sector-Transformation_Nov-2017-3.pdf. Accessed on 03-10-2019.

Central Statistics Office, Ministry of Statistics and Programme Implementation, Government of India, 2018. Energy statistics. http://mospi.nic.in/sites/default/files/publication_reports/Energy_Statistics_2018.pdf. Accessed on 12-07-2019.

Chandrayan, A., Bandyopadhyay, S. 2014. Cost optimal segregated targeting for resource allocation networks. *Clean Technologies and Environmental Policy*, 16(3), 455–465.

Chazen Global Insights, 2016. Why India is the fastest-growing economy on the planet. Economics & policy, ideas & insights. Columbia Business School. www8.gsb.columbia.edu/articles/chazen-global-insights/why-india-fastest-growing-economy-planet. Accessed on 12-07-2019.

Das, S. 2009. Let there be light. The Telegraph. www.telegraphindia.com/states/west-bengal/let-there-be-light/cid/1265015. Accessed on 12-07-2019.

Dubash, N. K., Rajan, S. C. 2001. The politics of power sector reform in India. http://pdf.wri.org/power_politics/india.pdf. Accessed on 12-07-2019.

Energy Efficiency Services Limited, 2019. About UJALA. https://eeslindia.org/content/raj/eesl/en/Programmes/UJALA/About-UJALA.html. Accessed on 12-07-2019.

Foo, D. C. Y., Tan, R. R., Ng, D. K. S. 2008. Carbon and footprint-constrained energy sector planning using cascade analysis technique. *Energy*, 33(10), 1480–1488.

Indian power sector.com, 2014. History of Indian power sector. http://indianpowersector.com/home/about/overview/. Accessed on 12-07-2019.

International Energy Agency, 2015. Nationally Determined Contribution (NDC) to the Paris agreement: India. www.iea.org/policiesandmeasures/pams/india/name-155210-en.php?s=dHlwZT1jYyZzdGF0dXM9T2s,&return=PG5hdiBpZD0iYnJlYWRjcnVtYiI-PGEgaHJlZj0iLyI-SG9tZTwvYT4gJnJhcXVvOyA8YSBocm VmPSIvcG9saWNpZXNhbmRtZWFzdXJlcy8iPlBvbGljaWVzIGFuZCBNZWFz dXJlczwvYT4gJnJhcXVvVv. Accessed on 05-08-2019.

Kale, S. S. 2014. *Electrifying India: Regional Political Economies of Development*. Stanford University Press.

Khosla, R., Dubash, N. K. 2015. What does India's INDC imply for the future of Indian electricity? Centre for Policy Research: The Climate Initiative Blog. https://cprclimateinitiative.wordpress.com/2015/10/15/what-does-indias-indc-imply-for-the-future-of-indian-electricity/. Accessed on 17-09-2019.

Krishna Priya, G. S. 2016. Carbon constrained power sector planning. Doctoral Dissertation. Indian Institute of Technology Bombay, India.

Krishna Priya, G. S., Bandyopadhyay, S. 2017. Multi-objective pinch analysis. *Resources, Conservation and Recycling*, 119, 128–141.

Ministry of Environment & Forests, Government of India, 2004. India's initial national communication to the United Nations framework convention on climate change. https://unfccc.int/resource/docs/natc/indnc1.pdf. Accessed on 12-07-2019.

Ministry of Law and Justice, Government of India, 2003. The electricity act, 2003. www.cercind.gov.in/Act-with-amendment.pdf. Accessed on 12-07-2019.

Ministry of New and Renewable Energy, Government of India, 2017. Annual report, 2016–2017. https://mnre.gov.in/file-manager/annual-report/2016-2017/EN/pdf/1.pdf. Accessed on 05-08-2019.

Ministry of Power, Government of India, 1998. Annual report, 1997–98. https://powermin.nic.in/sites/default/files/uploads/ar97-98.pdf. Accessed on 12-07-2019.

Ministry of Power, Government of India, 2000. Annual report, 1999–2000. https://powermin.nic.in/sites/default/files/uploads/Annual_Report_1999-00_English.pdf. Accessed on 12-07-2019.

Ministry of Power, Government of India, 2010. Annual report, 2009–2010. https://powermin.nic.in/sites/default/files/uploads/Annual_Report_2009-10_English.pdf. Accessed on 12-07-2019.

Ministry of Power, Government of India, 2014. Office memorandum, Deendayal Upadhyaya Gram Jyoti Yojana. https://powermin.nic.in/sites/default/files/uploads/Deendayal_Upadhyaya_Gram_Jyoti_Yojana.pdf. Accessed on 12-07-2019.

Ministry of Power, Government of India, 2015. Office memorandum, UDAY (Ujwal Discom Assurance Yojana) scheme for operational and financial turnaround of power distribution companies. https://powermin.nic.in/pdf/Uday_Ujjawal_Scheme_for_Operational_and_financial_Turnaround_of_power_distribution_companies.pdf. Accessed on 12-07-2019.

Ministry of Power, Government of India, 2017. Office memorandum, Pradhan Mantri Sahaj Bijli Har Ghar Yojana. https://powermin.nic.in/sites/default/files/webform/notices/OM_SAUBHAGYA_SIGNED_COPY.pdf. Accessed on 12-07-2019.

Ministry of Power, Government of India, 2018. Annual report, 2017–2018. https://powermin.nic.in/sites/default/files/uploads/MOP_Annual_Report_Eng_2017-18.pdf. Accessed on 12-07-2019.

Ministry of Power, Government of India, 2019a. Annual report year wise. https://powermin.nic.in/en/content/annual-reports-year-wise-ministry. Accessed on 12-07-2019.

Ministry of Power, Government of India, 2019b. Power sector at a glance ALL INDIA. https://powermin.nic.in/en/content/power-sector-glance-all-india. Accessed on 12-07-2019.

Ministry of Power, Government of India, 2019c. Executive summary on power sector, March–19. Central Electricity Authority. http://cea.nic.in/reports/monthly/executivesummary/2019/exe_summary-03.pdf. Accessed on 17-09-2019.

Patole, M., Bandyopadhyay, S., Foo, D. C., Tan, R. R. 2017. Energy sector planning using multiple-index pinch analysis. *Clean Technologies and Environmental Policy*, 19(7), 1967–1975.

Shenoy, U. V., Bandyopadhyay, S. 2007. Targeting for multiple resources. *Industrial & Engineering Chemistry Research*, 46(11), 3698–3708.

The World Bank, 2019. CO_2 emissions (metric tons per capita). https://data.worldbank.org/indicator/EN.ATM.CO2E.PC?locations=IN&most_recent_year_desc=false. Accessed on 12-07-2019.

United Nations, 1992. United Nations framework convention on climate change. http://unfccc.int/files/essential_background/background_publications_htmlpdf/application/pdf/conveng.pdf. Accessed on 12-07-2019.

United Nations Census Bureau Current Population, 2019. www.census.gov/popclock/print.php?component=counter. Accessed on 12-07-2019.

WorldData.info, Energy consumption in India. www.worlddata.info/asia/india/energy-consumption.php. Accessed on 12-07-2019.

8

Multi-Period Power Generation Planning for the United Arab Emirates (UAE)

Xiao Yien Lim and Shivakumar Tarun

Heriot-Watt University Dubai Campus

United Arab Emirates (UAE) is a country made up of seven emirates, namely Abu Dhabi, Dubai, Sharjah, Umm al-Quwain, Ajman, Ras Al Khaimah, and Fujairah. Located at the Middle East region which is well known for their huge reserves of fossil fuels, UAE is currently the world's eighth largest in proven reserves for oil and natural gas (BP, 2018). Due to its fast economic and population growth in recent years, the demand of energy in UAE is climbing rapidly. UAE power generation comes largely from natural gas, and this contributes to a large portion of the country's CO_2 emissions. Part of this energy is also used to produce potable water due to the scarcity of freshwater in this region. The challenges faced by UAE would be more complicated in view of the interlinkage between power and water. Balancing sustainability growth of the nation's development and the risk on climate change is the focus of this chapter. A multi-period planning strategy through 2050 is adopted for UAE power generation.

8.1 Overview of UAE Energy System and Future Energy Plans

More than 98% of the electricity supplied in the UAE is powered by gas and the remaining by oil (UAE Ministry of Energy & Industry, 2016). UAE is ranked the 14th highest energy and electric power uses per capita globally with a record of 11 GWh/capita annually, compared with world average of 3 GWh/capita (World Bank, 2019). The country's electricity generation capacity has been expanded significantly since the past 10 years, with 28.7 MW installed as of 2016 to meet the electricity peak demand; the latter normally happens during the summer period (UAE Ministry of Energy & Industry, 2016). According to the International Energy Agency, UAE carbon emission has recorded a tremendous increase of 63% between 2000 and 2010. If this trend is not curbed instantly, this would endanger the entire ecosystem of the country as well as the region (UAE Ministry of Energy & Industry, 2015).

Figure 8.1 shows the breakdown of greenhouse gas (GHG) emission in UAE, with the largest emission coming from the energy sector, while agriculture contributes the least. Within the energy sector, electricity and heat generation dominate around 41% of the carbon emission, with the remaining contributed by transport, oil and gas refining, manufacturing industries, and others (UAE Ministry of Energy & Industry, 2018). On the other hand, fossil fuels provide vital revenue to the country's economy through oil and gas exports. Since 2010, UAE has become a net-importer of gas as the country's demand began to outstrip its domestic gas production. This trend is similar to other nearby Gulf Cooperation Council (GCC) countries including Kuwait and Bahrain, due to domestic gas shortage (IRENA, 2015; Eveloy and Gebreegziabher, 2019). This has resulted in the fear of volatile gas prices and associated energy security concerns of the countries aforementioned.

The finiteness of the fossil fuels and the ambition to cut down CO_2 emissions of the country have driven UAE to diversify its energy resources. This has led to the national clean energy targets of 27% by 2021, and an increased target of 50% by 2050 (UAE Ministry of Energy & Industry, 2017, 2018). In this regard, the clean energy goals would be mainly supported by solar and nuclear powers. The first-ever nuclear plant in the region is expected to be completed by year 2020, with a total power output of 5.6 GWh from four reactors to meet 25% of the country's electricity demand (Treyer and Bauer, 2016; UAE Ministry of Energy & Industry, 2016). With falling prices of the solar panels and abundant solar radiation, solar power production will

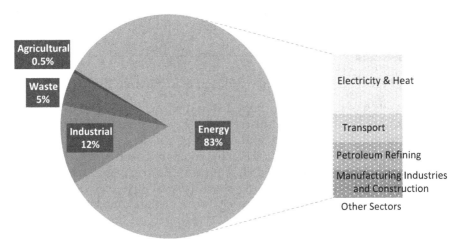

FIGURE 8.1
CO_2 emission of UAE based on sub-sectors (UAE Ministry of Energy & Industry, 2018).

be dominating the clean energy portfolio of the country up to 2050. Recent reports have indicated that a threefold of installation capacity are currently under construction and reviewed, which includes the 1.2 GW Barjeel project of Ras-Al-Khaimah, Mohammad Bin Rashid Al Maktoum Solar Park with a total capacity of 5,000 MW and the 1,177 MW of Noor Abu Dhabi (The National, 2019; Dubai Electricity and Water Authority, 2019; Khaleej Times, 2019). The UAE clean energy goals are considered realistic in addressing the climate change as well as being economically feasible, since the intermittent solar resources can be overcome and supported by the existing (combined cycle) gas power plants to ensure power stability and reliable supply to the grid (Treyer and Bauer, 2016).

8.2 Multi-Period Carbon-Constrained Energy Planning 🖳

Considering the UAE's present and projected energy, economic and environmental context, a conceptual future UAE power generation scenarios from 2021 to 2050 is showcased herein using the *multi-period* ATM (Ooi et al., 2014). This work is an extension of the CEPA of the UAE power generation sector by Lim et al. (2018). Energy sources for the indicated years are tabulated in Table 8.1, which comprises of natural gas, renewables, nuclear energy and coal with CCS integration. The selection of such potential future electricity mix scenarios are built based on information from various recent publications and statistics of the UAE (Eveloy and Gebreegziabher, 2019; Treyer and Bauer, 2016; Lim et al., 2018; UAE Ministry of Energy & Industry, 2016).

It is worth mentioning that three time intervals T_t are selected for this case study, with a period of 10 years each, i.e., T_1 (2021–2030), T_2 (2031–2040), and T_3 (2041–2050). The intermediate emission limits C_{LT} are assumed as 115, 125, and 125 Mt/y at the end of each time interval, consistency with the previous publication to have a reduction in emission to 2012 level of 62 Mt CO_2-e (Lim et al., 2018). It is also reasonable to have a higher reduction in the later years with respect to the learning curve of CCS technology and efficiency. The 2.4 GW new clean coal power plant will be incorporated with CCS when its operation began running in 2030; hence, its CO_2 intensity is set to 0.22 kg CO_2/kWh (Allen, 2011; UAE Ministry of Energy & Industry, 2016). The emission factor for nuclear is fixed at 0.01 kg CO_2/kWh to account for minor GHG releases from such systems. Owing to the limited alternative renewable resource potential and assuming no further expansion of planned nuclear and coal installation based on existing UAE plans, gas power plant would have to be retrofitted with CCS in order to meet the emission limit. A *removal ratio* (RR) of 0.80 and energy loss ratio (X) of 0.15 are selected (Tan et al., 2009).

TABLE 8.1

Energy Generation and CO_2 Emission Data of UAE Case Study

Energy Source	Fuel, $F_{S,i}$ (TWh/y)	CO_2 Intensity, $C_{S,i}$ (Mt/TWh)	CO_2 Emission, $E_{S,i}$ (Mt/y)
T_1 (2021–2030)			
Renewables	16.0	0	0.0
Nuclear	45.2	0.01	0.5
Coal-CCS	18.6	0.22	4.1
Gas	186.3	0.61	113.6
Total	266.1		118.2
T_2 (2031–2040)			
Renewables	95.5	0	0.0
Nuclear	45.5	0.01	0.5
Coal-CCS	18.2	0.22	4.0
Gas	295.4	0.61	180.0
Total	454.5		184.7
T_3 (2041–2050)			
Renewables	341.6	0	0.0
Nuclear	46.6	0.01	0.5
Coal-CCS	18.6	0.22	4.1
Gas	369.5	0.61	225.4
Total	776.4		230.0

The basics of ATM for targeting carbon capture can be found in Chapter 4 of this book. In this chapter, the ATM is set to minimize the overall compensatory power, F_{COMP} (see Equation 8.1), due to power losses during the retrofit of carbon capture across the entire planning horizon (i.e., T_1 to T_3).

$$\text{Minimum } F_{COMP} = \sum_t \sum_i F_{\text{Ret},i,t}\, X_i \qquad (8.1)$$

The constraints in solving the ATM model are given as follows.

The original energy output of the sources ($F_{S,i}$) and those to be retrofitted ($F_{\text{Ret},i}$) are located at their respectively intensity levels. Constraints are added to ensure that the amount of sources to be retrofitted will take non-negative values (Equation 8.2) and are bound to their maximum output values (Equation 8.3):

$$F_{\text{Ret},i,t} \geq 0 \quad \forall i \forall t \qquad (8.2)$$

$$F_{\text{Ret},i,t} \leq F_{S,i,t} \quad \forall i \forall t \qquad (8.3)$$

To ensure the retrofitting takes place in a logical time sequence, this is given in Equation 8.4:

$$F_{Ret,i,t+1} \geq F_{Ret,i,t} \quad \forall i \forall t \tag{8.4}$$

Taking into account the CCS retrofit, the net energy for each level k ($F_{Net,k,t}$) can be calculated using Equations 8.5 and 8.6 as follows:

$$F_{NS,k,t} = \begin{cases} F_{S,i,t} + F_{Ret,i,t}(1 - X_i) \\ F_{S,i,t} - F_{Ret,i,t} \end{cases} \tag{8.5}$$

$$F_{Net,k,t} = F_{NS,k,t} - F_{P,t} \quad \forall k \forall t \tag{8.6}$$

The *power cascade* is formed using Equation 8.7:

$$\delta_{k,t} = \begin{cases} 0 & k = 0 \\ \delta_{k-1,t} + F_{Net,k,t} & k \geq 1 \end{cases} \quad \forall t \tag{8.7}$$

The CO_2 *cascade* is formed using Equations 8.8 and 8.9:

$$\varepsilon_{k,t} = \begin{cases} 0 & k = 1 \\ \varepsilon_{k-1,t} + \delta_{k-1,t} \Delta C_{k-1,t} & k \geq 2 \end{cases} \quad \forall t \tag{8.8}$$

$$\varepsilon_{k,t} \geq 0 \quad \forall k \forall t \tag{8.9}$$

To achieve the CO_2 emission limit of 2012 level (62 Mt CO_2-e) by 2050, a reduction of 167.8 Mt/y of CO_2-e (= 230–62 Mt CO_2-e) has to be achieved. The model is solved with objective in Equation 8.1, subject to the constraints in Equations 8.2–8.9. The ATM in Figure 8.2 shows the result of CCS retrofit planning scenarios in gas power plants, representative of the 2021–2050 period based on the pre-set emission limit. The overall compensatory power (F_{COMP}) to supplement the energy losses during CCS retrofit is determined as 51.6 TWh/y, which were contributed by those in intervals T_1 (0.9 TWh/y), T_2 (18.3 TWh/y), and T_3 (32.3 TWh/y). In interval T_1, 6.5 TWh/y of power produced in the gas power plant is to be retrofitted with CCS, and this has increased to 122.3 and 215.1 TWh/y, in intervals T_2 and T_3, respectively. The corresponding retrofitting requirement has arisen from 3% in T_1 to 41% in T_2 and 58% in T_3, as more CO_2 is emitted as a result of larger power generation in the latter years.

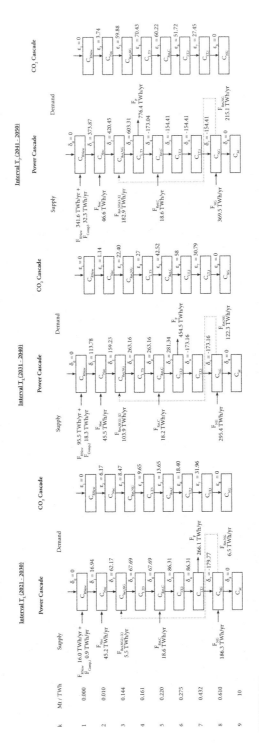

FIGURE 8.2

ATM results for UAE carbon-constrained power generation scenario.

8.3 Consideration of Energy-Water Nexus 🖳

Due to the arid climate and large energy inputs needed for desalination process, water and power are tightly interlinked in the UAE. Desalinated water is used for domestic and industrial needs while groundwater is mostly used in the agricultural sector (Treyer and Bauer, 2016). The energy required to desalinate water is proportional to the salinity of the source water and the technology employed. The common desalination techniques applied in UAE are multi-stage flash (MSF), multi-effect distillation (MED), and reverse osmosis (RO). The first two methods are thermal-based processes, in which energy is used for evaporation and condensation of water, hence require both heat and electricity. On the other hand, RO consumes only electrical energy in which a thin membrane is used to allow water to permeate from the saline side under high pressure. More than 80% of the potable water in UAE generates through thermal processes as they use the waste heat from the electricity generation process and hence are connected to power plants (IRENA, 2015).

Figure 8.3 shows a schematic diagram of a combined cycle power and water cogeneration plant which is extensively employed in the UAE. Combined cycle links the two power cycles (Brayton and Rankine) together such that the waste heat from the first cycle is used as the input for the second cycle in order to improve the overall thermal efficiency. The low-pressure steam output from the turbine can also subsequently be utilized for low-pressure

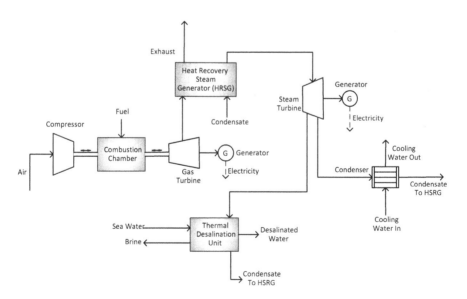

FIGURE 8.3
Combined cycle electricity and desalinated water cogeneration plant (Shahzad et al., 2019).

processes in the saltwater-desalination cycle. Siddiqi and Anadon estimated that UAE uses about 22% of its total electrical energy for desalination purpose (Siddiqi and Anadon, 2011).

Water is consumed in electricity generation. It was reported that global water footprint of electricity and heat production has increased significantly over the past years, mainly due to the increase in total energy demand and utilization of biomass as an energy source (Griffiths, 2017; Mekonnen et al., 2015). This trend is expected to rise continuously, putting additional pressure on the limited freshwater resources, especially in the GCC region in which freshwater supply is scarce. Water requirement in power production can vary substantially, depends on the fuel type, generation technology as well as the cooling systems, as shown in Table 8.2. Renewable energy sources like wind, solar use relatively less water than fossil-fired and nuclear power plants. Advanced fossil fuel technologies such as combined cycle have lower water intensity due to higher thermal efficiency, while CCS retrofit incurs water footprint penalties due to the fact that CO_2 absorption requires large amount of cooling water (Fricko et al., 2016; Lim et al., 2018). Using the ATM result in Figure 8.2, the water demand of each interval, W_t of UAE power generation is calculated using Equation 8.10.

$$W_t = \sum_i W_i F_{\text{Ret},i}(1-X_i) + W_i F_{\text{S},i} \qquad \forall t \qquad (8.10)$$

where W_i represents the water intensity for each of the fuel source.

Accordingly, the water required for power generation increases from $232\,km^3$ followed by 374 to 506 km^3 for intervals T_1, T_2, and T_3, respectively, as more CCS technology is integrated into gas power plant to meet the CO_2 limit.

Currently, UAE is still overwhelmingly dependent on natural gas cogeneration to produce electricity and desalinated water. With its growing population, increasing water and energy demands, and decreasing freshwater supplies, it is becoming more vulnerable to the challenges that the energy-water nexus presents in the future. Shifting to a low-carbon emission power sector in UAE will deeply affect the water sector, as less electricity generated

TABLE 8.2

Water Intensity Data for Power Generation (Lim et al., 2018)

	Water Intensity, W_i (km³/TWh)
Renewables	0.1
Nuclear	2.5
Coal-CCS	2.0
Gas	0.4
Gas-CCS	1.4

from thermal plants will lead to less waste heat and hence reduces the desalinated water production. This water gap would need to be covered either by new RO plants, or through new MED plants that use low-grade heat from solar or geothermal (IRENA, 2015; UAE Ministry of Energy & Industry, 2016). With more renewable energy alongside with nuclear and CCS retrofitted plants playing the future energy role, it is important to incorporate a water-constrained planning for sustainable power generation in the UAE.

8.4 Multi-Period Consideration for Carbon- and Water-Constrained Energy Planning

With regards to the strong energy-water linkage in the UAE, an extension of the power sector planning incorporating water constraint is demonstrated in Figure 8.4 on the basis that the CO_2 load is allowed to transfer among the time intervals while achieving the same targeted CO_2 emission limit at the end of the planning horizon, i.e., year 2050. Therefore, Equation 8.8 is revised as Equation 8.11 that follows:

$$\varepsilon_{k,t} = \begin{cases} 0 & k = 1 \\ \varepsilon_{k-1,t} + \delta_{k-1,t}\Delta C_{k-1,t} \pm CB_{k,t-1} & k \geq 2 \end{cases} \quad \forall t \qquad (8.11)$$

where $CB_{k,t-1}$ indicates transferred carbon load across the time intervals. In this case, a positive value (+) indicates the transfer of carbon load from earlier time to later time interval, while negative (–) shows the transfer in the reverse direction.

In this scenario, the total water consumption of power generation is limited to a maximum of 30% of the previous time interval, in order to conserve water in this oil abundant but water scarce country (see Equation 8.12). This change in water requirement can be assumed to have a negligible effect on the power demand.

$$W_t \leq 1.3 \times W_{t+1} \quad \forall t \qquad (8.12)$$

The ATM is solved with objective in Equation 8.1 and subject to the constraints in Equations 8.2–8.7 and 8.9–8.12, with results shown in Figure 8.4. For this case, the overall compensatory power (F_{COMP}) of 51.6 TWh/y was contributed by those in intervals T_1 (9.9 TWh/y), T_2 (16.0 TWh/y), and T_3 (25.7 TWh/y). Figure 8.4 also shows that larger amount of retrofit is taken place in interval T_1, i.e., 66 TWh/y as compared to 6.5 TWh/y in the earlier scenario when water constraint was not considered (Figure 8.2). In T_1, the CO_2 emission is recorded at 86 Mt/y which is below the permitted level of 115 Mt/y

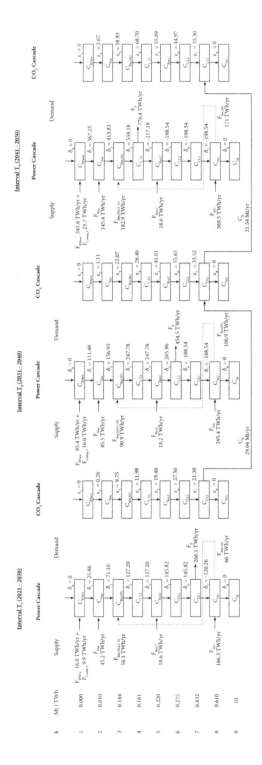

FIGURE 8.4
ATM results for consideration of water-energy nexus in UAE power generation sector.

(see Figure 8.5), with a corresponding water usage of 278.6 km³. An opposite trend was found in both intervals T_2 and T_3, where less water is required in the power sector while the carbon emission has increased to 132.5 Mt/y (T_2) and 146.5 Mt/y (T_3) due to more stringent water resource than carbon limit (see Figure 8.5). At such, 106.9 TWh/y of power produced in the gas plant is to be retrofitted with CCS in T_2 and a marginal increase to 171 TWh/y in T_3. The water usage is calculated as 362.1 km³ for T_2 and 470.8 km³ for T_3, a marginal increase of 2% while a drop of 13%, respectively from those in Figure 8.2, where only CO_2 emission is taken as the main concern. In other words, the carbon load debts of 7.5 Mt/y in T_2 and 21.5 Mt/y in T_3 are supplemented by the carbon credits of 29 Mt/y (from T_1) in this optimal strategy. Hence, the resulting distribution share of the CCS retrofitted plants throughout the entire planning duration has changed to 35% in T_1, 36% in T_2, and 46% in T_3, a more gradual increment in the retrofitting plan in comparison to the carbon-constrained scenario.

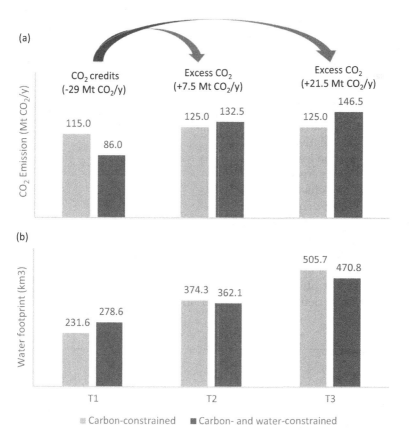

FIGURE 8.5
Illustration of (a) CO_2 emission and (b) water footprint comparison for T_1 to T_3 intervals.

8.5 Conclusion

UAE has set a 2050 national energy plan to move from a traditional fuel-based to energy sources with inherently low carbon footprints that includes solar and nuclear to meet the environmental goal. In order to effectively reduce the emissions of GHGs, apart from adopting nuclear and renewable technologies, a parallel implementation of CCS for natural gas plants is necessary. The carbon-constrained power planning discussed in Section 8.2 shows a large scale of CCS retrofit (more than 90%) to be implemented between 2031 and 2050, i.e., intervals T_2 and T_3. Nonetheless, integration of CCS into the power generation system has important implications on the regional water use and management. Hence, taking into consideration the water constraint into the same context as described in Section 8.4, a more evenly distributed CCS retrofitting profile is observed, where 30% takes place in the first interval by 2030, and increases to 35% each in the latter two intervals. This case study optimizes the deployment of CCS retrofit in the planning timeline, while simultaneously considering water constraints and relevant energy sources available.

Nomenclature

δ	Cumulative energy
ε	Cumulative CO_2 emission
CB	Carbon load transfer
C_S	CO_2 intensity
F_{COMP}	Compensatory power
F_P	Power demand
F_{Ret}	Fuel sources that need to be retrofitted
F_S	Fuel source
RR	Removal ratio
X	Energy loss ratio
W_i	Water intensity
W_t	Water demand of time interval

References

Allen, S. 2011. Carbon footprint of electricity generation. POSTnote Update, no. 383, pp. 1–4.
BP, 2018. *BP Statistical Review of World Energy*, 67th Ed. www.bp.com/content/dam/bp/business-sites/en/global/corporate/pdfs/energy-economics/statistical-review/bp-stats-review-2018-full-report.pdf. Accessed July 2019.

Dubai Electricity and Water Authority, 2019. Mohammed Bin Rashid Al Maktoum solar park - A leading project that promotes sustainability in the UAE. www.dewa. gov.ae/en/about-dewa/news-and-media/press-and-news/latest-news/2019/03/ mohammed-bin-rashid-al-maktoum-solar-park. Accessed July 2019.

Eveloy, V., Gebreegziabher, T. 2019. Excess electricity and power-to-gas storage potential in the future renewable-based power generation sector in the United Arab Emirates. *Energy*, 166, 426–450.

Fricko, O., Parkinson, S. C., Johnson, N., Strubegger, M., Van Vliet, M. T. H., Riahi, K. 2016. Energy sector water use implications of a 2°C climate policy. *Environmental Research Letters*, 11(3), 034011.

Griffiths, S. 2017. A review and assessment of energy policy in the Middle East and North Africa Region. *Energy Policy*, 102(December 2016), 249–269.

IRENA, 2015. *Renewable Energy Prospects: United Arab Emirates, REmap 2030 Analysis.* IRENA, Abu Dhabi.

Khaleej Times, 2019. World's largest solar project begins operation in UAE - Khaleej Times. Abu Dhabi. www.khaleejtimes.com/business/energy/worlds-largest-solar-project-begins-operation-in-uae. Accessed July 2019.

Lim, X. Y., Foo, D. C., Tan, R. R. 2018. Pinch analysis for the planning of power generation sector in the United Arab Emirates: A climate-energy-water nexus study. *Journal of Cleaner Production*, 180, 11–19.

Mekonnen, M. M, Gerbens-Leenes, P. W., Hoekstra, A. Y. 2015. The consumptive water footprint of electricity and heat: A global assessment. *Environmental Science: Water Research & Technology*, 1(3), 285–297.

Ooi, R. E., Foo, D. C., Tan, R. R. 2014. Targeting for carbon sequestration retrofit planning in the power generation sector for multi-period problems. *Applied Energy*, 113, 477–487.

Shahzad, M. W., Burhan, M., Ybyraiymkul, D., Ng, K. C. 2019. Desalination processes' efficiency and future roadmap. *Entropy*, 21(1), 84.

Siddiqi, A., Anadon, L. D. 2011. The water-energy nexus in Middle East and North Africa. *Energy Policy*, 39, 4529–4540.

Tan, R. R., Ng, D. K. S., Foo, D. C. Y. 2009. Pinch analysis approach to carbon-constrained planning for sustainable power generation. *Journal of Cleaner Production*, 17(10), 940–944.

The National, 2019. Ras Al Khaimah aims to add 1.2GW of renewables to grid by 2040. www.khaleejtimes.com/business/energy/worlds-largest-solar-project-begins-operation-in-uae. Accessed July 2019.

Treyer, K., Bauer, C. 2016. The environmental footprint of UAE's electricity sector: Combining life cycle assessment and scenario modeling. *Renewable and Sustainable Energy Reviews*, 55, 1234–1247.

UAE Ministry of Energy & Industry, 2015. The UAE state of energy report 2015.

UAE Ministry of Energy & Industry, 2016. The UAE state of energy report 2016.

UAE Ministry of Energy & Industry, 2017. The UAE state of energy report 2017.

UAE Ministry of Energy & Industry, 2018. United Arab emirates fourth national communication report.

World Bank, 2019. Electric Power Consumption (KWh per Capita) | data. The World Bank - Data. https://data.worldbank.org/indicator/EG.USE.ELEC.KH.PC. Accessed July 2019.

9

Carbon Emission Reduction Targeting for Bioenergy Supply Chain in Malaysia

Viknesh Andiappan and Lik Yin Ng
Heriot-Watt University Malaysia

Dominic C. Y. Foo
The University of Nottingham Malaysia

9.1 Introduction

During the COP21 in 2015, Malaysia pledged to reduce its greenhouse gases emission intensity per unit gross domestic product by 45% in 2030 from the levels in 2005 (UN Framework Convention on Climate Change, 2015). To initiate efforts in meeting this pledge, Malaysia launched the National Renewable Energy Policy and Action Plan (SEDA, 2018). The objective of this policy was to increase the share of renewable energy resource in the generation of electricity generation. However, the outcome of general elections in 2018 brought change in Malaysian government agencies and policymakers. Since then, the appointed Minister of Energy, Technology, Science, Environment, and Climate Change has formulated new initiatives to encourage the implementation of renewable energy (MESTECC, 2019). Based on this, it is clear that carbon intensity reduction and the future of the renewable energy sector in Malaysia remains a key focus (Brown, 2018).

Malaysia would require systematic planning of its renewable energy sector to cut down the national carbon intensity. CEPA (Tan and Foo, 2007) is a systematic tool that can be used to assess and plan carbon reduction targets in Malaysia. On top of this, results from CEPA can provide insights on how a bioenergy supply chain (BSC) can be planned. As such, this chapter couples the benefits of both CEPA and BSC planning under one methodology. To perform supply chain planning, this chapter uses a mathematical programming method based on superstructure approach. The latter is a method used widely in areas such as design of utility systems (Yeomans and Grossmann, 1999), eco-industrial parks (Andiappan et al., 2016), and biorefineries (Ng et al., 2015).[1] Superstructural optimization is used to provide all potential routes within a BSC where biomass may be transported for energy conversion. Based on these routes, a mathematical model is developed. The mathematical model is then maximized or minimized based on a given objective to determine the optimum route(s) (Andiappan, 2017). These objectives can include (but not limited to) economic performance and environmental impact of the BSC. In this chapter, the objective is to determine the optimal BSC with minimum total annualized cost (TAC).

9.2 Problem Statement

Selangor is one of the most developed states in Malaysia. Being a highly developed state, Selangor requires high amount of electricity for its population. However, this high electricity demands are currently met via power plants powered by fossil fuels, which in turn generate high carbon emissions. Table 9.1 shows the electricity generation share in Selangor based on fuel types. The aim of this chapter is to evaluate the potential of carbon emission reductions in which a palm oil-based BSC would be able to attain by displacing a portion of the fossil-based electricity in Selangor. This BSC is expected to collect biomass generated from seven mills (and their respective plantations), in order to generate electricity adjacent to a local substation. The locations of these mills and their distances from the substation are shown in Figure 9.1 and Table 9.2, respectively. The mills and plantations both generate biomass as waste from their operations. These biomass wastes include oil palm trunk (OPT), oil palm fronds (OPF), empty fruit bunches (EFB), palm kernel shell (PKS), palm mesocarp fiber (PMF), and wastewater known as palm oil mill effluent (POME) (Sadhukhan et al., 2018). The amount of biomass wastes generated by each plantation is summarized in Table 9.3 (Leong et al., 2019).

To determine the optimal route for electricity generation, a BSC superstructure is developed (see Figure 9.2). As shown, Figure 9.2 considers the decision of investing in several electricity generation technologies, i.e., gasifiers and combined heat and power (CHP) systems. Aside from that, the EFB pathway has several options for pre-treatment, which are hot air drying, superheated steam drying, and size reduction systems.

---

[1] See Chapter 5 for detailed discussion of this technique for carbon-constrained energy planning.

TABLE 9.1

Electricity Generation Share in Selangor (Energy Commission of Malaysia, 2017)

Type of Power Plant	Output (TJ)	Carbon Emission Factor (million t/TJ)
Natural gas	60,176	51.5
Coal	22,314	105.0

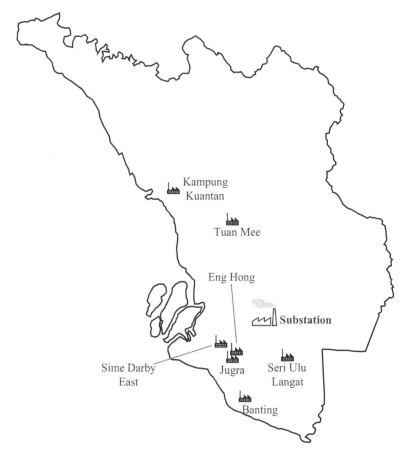

FIGURE 9.1
Location of palm oil mills and plantations (Leong et al., 2019).

The data used for this chapter (technology efficiencies, energy require-ments, cost data, etc.) were reported in Leong et al. (2019) (see Tables 9.4–9.7). The methodology used to determine the optimal BSC (based on CEPA targets) is discussed further in the following section.

TABLE 9.2

Distances of Mills and Plantations from Electrical Substation
(Leong et al., 2019)

Mills and Plantations	Distance to Substation (km)
Jugra	28
Eng Hong	30
Sime Darby East	24
Banting	50
Kampung Kuantan	77
Tuan Mee	42
Seri Ulu Langat	32

TABLE 9.3

Amount of Biomass Available Based on Total Area of Plantation (Leong et al., 2019)

Location	Area (ha)	EFB (t/h)	PMF (t/h)	PKS (t/h)	POME (m³/h)	OPT (t/h)	OPF (t/h)
Jugra	6.29	13.49	8.84	3.23	4.1	236	78
Eng Hong	5.03	10.79	7.07	2.58	3.3	189	63
Sime Darby East	5.66	12.14	7.96	2.91	3.7	212	70
Banting	2.52	12.14	7.96	2.91	3.7	212	70
Kampung Kuantan	5.66	5.40	3.54	1.29	1.6	94	31
Tuan Mee	2.52	7.28	4.77	1.74	2.2	127	42
Seri Ulu Langat	3.40	12.14	7.96	2.91	3.7	212	70

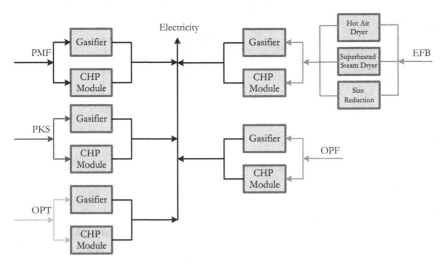

FIGURE 9.2
BSC superstructure (Leong et al., 2019).

TABLE 9.4

Technical Considerations for Technologies (Leong et al., 2019)

Technology	Data	Refs
Hot air drying	• Moisture content of EFB before and after the dryer are 60% and 10%, respectively • Efficiency of 85.93% • Requires 2.31 MJ/kg biomass from PP or CHP	Aziz et al. (2011)
Superheated steam drying	• Moisture content of EFB before and after the dryer are 60% and 10%, respectively • Efficiency of 86.72% • Requires 2.18 MJ/kg biomass from PP or CHP	Aziz et al. (2011)
PP	• Efficiency is 40% • Maximum power produced by the PP is 419.4 TJ	Loh (2017)
CHP	• Efficiency is 48.9% (with 26.5% for heat and 22.4% for electricity) • Maximum power produced by the PP is 1,175.94 TJ	Loh (2017)
Gasifier	• Maximum weight of biomass waste handled is 1,612 t	Forbes International Co. LTD (2018)

TABLE 9.5

Technical Considerations for Supply Chain (Leong et al., 2019)

Biomass transportation costs	0.23 USD t km^{-1} traveled (Zafar, 2018)

TABLE 9.6

Calorific Values of Biomass before and after Incorporating Equipment Efficiency (Leong et al., 2019)

Pathway	Actual CV (MJ/kg)	Heat Loss for Drying (MJ/kg)	Final CV for PP (MJ/kg)	Final CV for CHP (MJ/kg or MJ/m³) Electricity	Steam	Refs
EFB (hot air drying)	17.81	2.31	4.814	3.99	2.41	Aziz et al. (2011)
EFB (superheated steam drying)	17.81	2.18	4.944	3.99	2.54	Aziz et al. (2011)
EFB (size reduction)	16.5	-	6.6	3.70	4.37	EPA (2018)
PKS	20.09	-	8.036	4.50	5.32	Paul et al. (2015)
PMF	14.51	-	5.804	3.25	3.85	Sarawak Energy (2018)
POME	22	-	8.8	4.93	5.83	Onoja et al. (2018)
OPT	17.47	-	6.988	3.91	4.63	Rashid et al. (2017)
OPF	5.15	-	2.06	1.15	1.36	EPA (2018)

CV, Calorific value

TABLE 9.7

Equipment Cost for Each Pathway (Leong et al., 2019)

Pathway	Fixed Cost (USD)	Variable Cost (USD/ton or USD/kW)	Refs
EFB (hot air drying)	446,455.75	0.0245	IRENA (2013)
EFB (superheated steam drying)	446,455.75	0.0245	IRENA (2013)
EFB (size reduction)	674,000	20	IRENA (2013)
PKS	14,250,000	3,700	Forbes International Co. LTD (2018)
PMF	14,250,000	3,700	Forbes International Co. LTD (2018)
POME	451,655	4,350	MP Energy (2018)
OPT	14,250,000	3,700	MP Energy (2018)
OPF	14,250,000	3,700	MP Energy (2018)
PP	18,000,000	211.6	Loh (2017)
CHP	11,780,000	7,670	Loh (2017)

9.3 Methodology

Figure 9.3 shows the general framework used to solve the problem stated in the previous section. As shown, the framework begins with CEPA. CEPA is used to determine a suitable carbon emission reduction target. Based on this target, the minimum required renewable energy is determined. Detailed steps of CEPA can be found in Chapter 2 of this book, as well as in Tan and Foo (2017). This value will then be used as input for the BSC superstructure optimization.

Prior to superstructure optimization, Monte Carlo simulation is performed. Monte Carlo simulation is a method that builds models of possible results for a given parameter that is uncertain. Essentially, it substitutes a range of values, each time using a different set of random values from the probability functions. By using probability distributions, variables can have different probabilities of outcomes occurring. Probability distributions are a much more realistic way of describing uncertainty in variables.

In this chapter, Monte Carlo simulation is performed via LINGO optimization software v14.0 (Lindo Systems Inc., 2018) to generate growth factors (G_{gh}) values for oil palm plantation, based on an assumed normal probability distribution with unique mean and standard deviation values. Alternatively, Monte Carlo simulation for these growth factors can be performed via Visual Basic Application (VBA) in Microsoft Excel. Growth factors represent the potential increase or decrease in yearly yield for resources h and represents

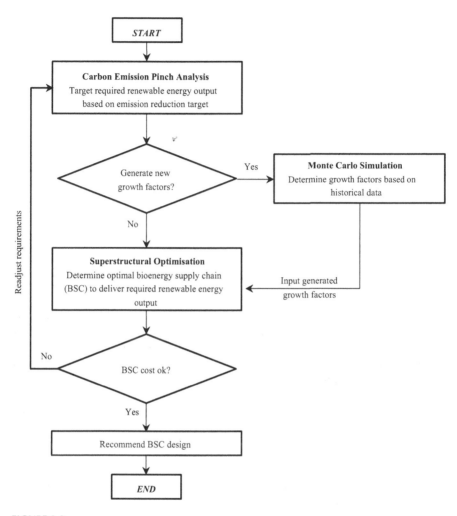

FIGURE 9.3
Methodology to determine optimal BSC (based on CEPA).

the performance of each plantation g. If there are possibilities of disruption (e.g., the water crisis, droughts), the growth factor can be set below unity. On the other hand, higher growth factor (more than unity) may also be set if the plantation yield is higher than its prior period.

The mean and standard deviation used in this chapter are set to 0.89 and 0.15, respectively. The growth factors generated from the Monte Carlo simulation are then used as input for the BSC superstructure optimization. In the latter, possible pathways are screened and the optimal BSC with minimum cost based on minimum renewable energy target (as determined by CEPA) is selected. The generic framework for the superstructure optimization in this chapter is described further in Figure 9.4.

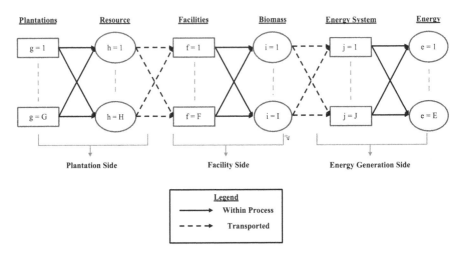

FIGURE 9.4
Generic superstructure for BSC optimization. (Modified from Leong et al., 2019.)

As shown, the generic superstructure in Figure 9.4 starts with plantations $g \in G$. In plantations $g \in G$, resources h are planted with an area A_{gh}. Resources here refer to crops that are harvested from the plantations and process in facilities. The processing of such resources later leads to the generation of biomass. The available amount of resource (i.e., crop) h, $F_{gh}^{Res,Av}$ is produced at a yield of Y_{gh} and growth factor G_{gh} as shown in Equation 9.1. Meanwhile, it is worth noting that the values set for G_{gh} would be generated by the Monte Carlo simulation mentioned previously.

$$F_{gh}^{Res,Av} = A_{gh}Y_{gh}G_{gh} \quad \forall g \forall h \tag{9.1}$$

The available resource h can be transported to another facility to undergo conversion or exported for further use, as shown in Equation 9.2. The total flow of resource h transported to facility f is given by F_{ghf}^{Trans}

$$F_{gh}^{Res,Av} \geq \sum_{f=1}^{F} F_{ghf}^{Trans} \quad \forall g \forall h \tag{9.2}$$

The flow of resource h transported to facility f to generate biomass i is as shown in Equation 9.3. The conversion of resource h to biomass i is given by V_{hfi}. The flow of biomass i produced at facility f is represented by F_i^{Bio}.

$$\sum_{g=1}^{G} \sum_{f=1}^{F} F_{ghf}^{Trans} V_{hfi} = F_i^{Bio} \quad \forall i \forall h \tag{9.3}$$

Following this, Equation 9.4 then shows the split of F_i^{Bio} further downstream. As shown, the flow F_i^{Bio} can be split for transport to an energy system j (F_{ij}).

$$F_i^{Bio} \geq \sum_{j=1}^{J} F_{ij} \quad \forall i \tag{9.4}$$

The flow of biomass i transported to energy system j is then converted to energy e as shown in Equation 9.5. The conversion of biomass i to energy e is given by V_{ije}. The amount of energy e produced at energy system j is given by F_{je}^{Energy}.

$$\sum_{i=1}^{I} F_{ij} V_{ije} = F_{je}^{Energy} \quad \forall e \forall j \tag{9.5}$$

It is important to note that the sum of F_{je}^{Energy} is the total renewable energy produced by the BSC. In this respect, the sum of F_{je}^{Energy} is equal to E^{RE}, which is the minimum renewable energy target determined by CEPA. In other words, Equation 9.6 represents the linkage between CEPA and the superstructure model.

$$E^{RE} = \sum_{j=1}^{J} \sum_{e=1}^{E} F_{je}^{Energy} \tag{9.6}$$

The annualized capital cost of the supply chain (CAP) is computed using Equation 9.7. CAP is determined based on the capacity of energy system j, which is calculated using the energy produced F_{je}^{Energy} and the variable cost per unit energy produced VC_j. CRF_j is the capital recovery factor for energy system j. Meanwhile, the fixed cost (FC_j) is activated using a binary integer b_j which denotes the existence of energy system j in the supply chain. b_j is governed by the following Equation 9.8, which shows the upper and lower capacities for a given energy system j.

$$CAP = \sum_{j=1}^{J} \sum_{e=1}^{E} \left(F_{je}^{Energy} VC_j + b_j \times FC_j \right) \times CRF_j \tag{9.7}$$

$$F_{ij}^{min} \times b_j \leq F_{ij} \leq F_{ij}^{max} \times b_j \tag{9.8}$$

Aside from this, the operational cost of the supply chain ($OPEX$) is computed using Equation 9.9. $OPEX$ is determined based on the transportation costs, which is calculated using the flow transported F_{ghf}^{Trans} and F_{ij}. The cost per unit flow transported is given by OFC_{ghf} and OFC_{ij}.

$$OPEX = \sum_{g=1}^{G}\sum_{h=1}^{H}\sum_{f=1}^{F}\left(F_{ghf}^{\text{Trans}}\text{OFC}_{ghf}\text{d}_{ghf}\right) + \sum_{i=1}^{I}\sum_{j=1}^{J}\left(F_{ij}\text{OFC}_{ij}\text{d}_{ij}\right) \qquad (9.9)$$

The TAC is given by the summation of capital expenditure *(CAPEX)* and *OPEX*, given as in Equation 9.10.

$$TAC = CAP + OPEX \qquad (9.10)$$

Note that the superstructure model in Equations 9.1–9.10 is a mixed integer linear program (MILP). It is also important to note that Equation 9.6 is the key constraint that links both CEPA and superstructure optimization. The superstructure optimization was performed using an optimization software LINGO v14.0 (Lindo Systems Inc., 2018).

9.4 Results and Discussion 💻

Figure 9.5 shows the EPPD from CEPA,[2] plotted using the data in Table 9.1. As shown, the dotted lines represent the status quo or base case. The dotted straight line represents the electricity demand in the state of Selangor, Malaysia. Meanwhile, the dotted curve shows the source curve from power plants fueled by natural gas and coal. The solid lines in Figure 9.5 are the results from using the methodology shown in Figure 9.4. Essentially, the solid lines represent the source curve that has been displaced due to the introduction of energy systems fueled by biomass. Such displacement reduced CO_2 emissions from 5.45 to 4.96 million t/y. This reduced CO_2 emissions corresponds to an intensity of 60.1 million t/TJ (= 4,960 kt/82,490 TJ) which essentially mounts to a CO_2 reduction of 8.95%. In order to achieve this CO_2 emission reduction, Figure 9.5 suggests that the minimum electricity output required from biomass fueled energy systems is 4,644 TJ.

Following this, Table 9.8 shows the growth factors generated randomly by the Monte Carlo simulation. Meanwhile, Figure 9.6 shows the probability distribution of growth factors generated by the Monte Carlo simulation. Alongside the 4,644 TJ target, these growth factors are then used as inputs for BSC superstructure model. The model was then solved by maximizing Equation 9.10 subject to Equations 9.1–9.9. The optimal value for the objective function was determined as 6.61 trillion USD/year and the optimal BSC is shown in Figure 9.7.

---

[2] See Chapter 2 for the detailed steps in generating the EPPD.

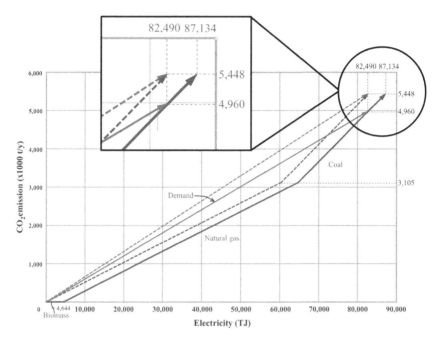

FIGURE 9.5
EPPD for case study (solid lines are the shifted composite curves that determine the minimum renewable energy (i.e., biomass) needed to achieve required emission reductions from base case shown in dotted lines) (Leong et al., 2019).

TABLE 9.8

Growth Factors Generated for Palm Oil Plantations by Monte Carlo Sampling

	Jugra	Eng Hong	Sime Darby East	Banting	Kampung Kuantan	Tuan Mee	Seri Ulu Langat
Growth factors	1	0.92	0.92	0.91	0.64	0.62	0.61

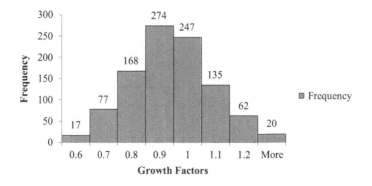

FIGURE 9.6
Probability distribution of growth factors generated by Monte Carlo simulation.

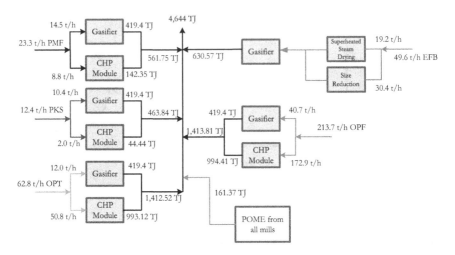

FIGURE 9.7
Optimal BSC design.

As shown in Figure 9.7, the optimal BSC contains a total energy output share of 2,308.17 TJ (= 419.4 + 419.4 + 419.4 + 419.4 + 630.57) from gasifier systems, 2,174.32 TJ (= 142.35 + 44.44 + 993.12 + 994.41) from CHP systems and 161.37 TJ generated from POME. It is observed that the portion of PMF and PKS channeled into gasifier systems were much larger compared to those entering the CHP systems. However, it is the opposite for OPT and OPF pathways, whereby a large portion of OPT and OPF was distributed to CHP systems. The reason for this is because the gasifier systems have higher efficiency to produce electricity. However, these gasifier systems are constrained by their respective maximum capacities. So, when the gasifier systems reach their maximum capacity, CHP systems were chosen to supply the remaining portion of electricity output targets. For the EFB pathway, superheated steam dryer and size reduction technologies were selected as pre-treatment stages prior to the gasifier. This is particularly due to the high initial moisture content in EFB, which is not suitable for the gasifier.

9.5 Conclusion

In this chapter, a hybrid methodology was demonstrated to systematically plan and optimize a palm-based BSC in Malaysian. This methodology first used CEPA to determine the minimum renewable energy requirement based on a given carbon emission reduction target of 8.95%. Based on this minimum requirement, a BSC superstructure is optimized to establish the optimal routes to achieve the set emission reduction target. Results suggest that the optimized BSC combined the use of technologies such as gasifiers, CHP modules, hot air dryers, and size reduction systems to produce 4,644 TJ electricity output.

Acknowledgement

The authors of this chapter would like to express their gratitude towards Prof. Kathleen Aviso for her kind technical support and advice.

Nomenclature

Indices

k	Index for levels
h	Index for resources
g	Index for plantations
i	Index for biomass
f	Index for facility/export
j	Index for energy system/export
e	Index for energy

Parameters

A_{gh}	Area of plantation g planted with resource h
Y_{gh}	Yield of resource h from plantation g
G_{gh}	Growth factor of resource h from plantation g
$F_{gh}^{Res,Av}$	Available amount of resource h from plantation g
V_{hfi}	Conversion of resource h to biomass i
V_{ije}	Conversion of biomass i to energy e
VC_j	Variable cost for energy system j per unit energy produced
CRF_j	Capital recovery factor for energy system j
FC_j	Fixed cost for energy system j per unit energy produced
OFC_{ghf}	Cost per unit flow of resource h transported from plantation g to facility f
OFC_{ij}	Cost per unit flow of biomass i transported to energy system j
d_{ghf}	Distance to transport resource h transported from plantation g to facility f
d_{ij}	Distance to transport resource i transported to energy system j

F_{ij}^{min} Minimum operating capacity for a given energy system j

F_{ij}^{max} Maximum operating capacity for a given energy system j

Variables

F_{ghf}^{Trans} Flow of resource h transported from plantation g to facility f

F_i^{Bio} Flow of biomass i produced

F_{ij} Flow of biomass i transported to energy system j

F_{je}^{Energy} Amount of energy e produced at energy system j

b_j Integer denoting existence of energy system j in the supply chain

CAP Annualized capital costs of the supply chain

$OPEX$ Operational costs of the supply chain

References

Andiappan, V., Tan, R. R., Ng, D. K. S. 2016. An optimization-based negotiation framework for energy systems in an eco-industrial park. *Journal of Cleaner Production*, 129(August), 496–507.

Andiappan, V. 2017. State-of-the-art review of mathematical optimisation approaches for synthesis of energy systems. *Process Integration and Optimization for Sustainability*, 1(3), 165–188.

Aziz, M. K. A., Morad, N. A., Wambeck, N., Shah, M. H. 2011. Optimizing palm biomass energy though size reduction. In *2011 4th International Conference on Modeling, Simulation and Applied Optimization, ICMSAO 2011*. doi:10.1109/ICMSAO.2011.5775516.

Brown, V. 2018. Yeo: You Don't Need to Know Me, You Need to Know How - Nation. *The Star Online*, July.

Energy Commission of Malaysia, 2017. Peninsular Malaysia electric supply outlook 2017. Suruhanjaya Tenaga (Energy Commission).

EPA. 2018. Biomass CHP catalog.

Forbes International Co. LTD, 2018. Palm Kernel Shell (PKS), renewable energy, renewable fuel, biofuel, biomass pellet, Palm Kernel Shell burner, energy management, environmentally fuel, green energy. www.fbs-world.com/en/product-2.html.

IRENA, 2013. Biomass-fired power generation operations and maintenance costs | Global CCS Institute. https://hub.globalccsinstitute.com/publications/renewable-power-generation-costs-2012-overview/83-biomass-fired-power-generation-operations-and-maintenance-costs.

Leong, H., Leong, H., Foo, D. C. Y., Ng, L. Y., Andiappan, V. 2019. Hybrid approach for carbon-constrained planning of bioenergy supply chain network. *Sustainable Production and Consumption*, 18, 250–267.

Lindo Systems Inc, 2018. LINGO: The modeling language and optimizer.

Loh, S. K. 2017. The potential of the Malaysian oil palm biomass as a renewable energy source. *Energy Conversion and Management*, 141, 285–298.

MESTECC, 2019. Initiatif MESTECC 2019.

MP Energy, 2018. Market info - Palm Kernel Shell. http://mpenergy.com.my/docs/market_info/palm-kernel-shell.pdf.

Ng, L. Y., Andiappan, V., Chemmangattuvalappil, N. G., Ng, D. K. S. 2015. Novel methodology for the synthesis of optimal biochemicals in integrated biore-fineries via inverse design techniques. *Industrial and Engineering Chemistry Research*, 54(21). doi:10.1021/acs.iecr.5b00217.

Onoja, E., Chandren, S., Razak, F. I. A., Mahat, N. A., Wahab, R. A. 2018. Oil palm (Elaeis Guineensis) biomass in Malaysia: The present and future prospects. *Waste and Biomass Valorization*, 10, 2099–2117.

Paul, O. U., John, I. H., Ndubuisi, I., Peter, A., Godspower, O. 2015. Calorific value of palm oil residues for energy utilisation. *IJEIR*. doi:10.1093/carcin/bgt085.

Rashid, S. R. M., Saleh, S., Samad, N. A. F. A. 2017. Proximate analysis and calorific value prediction using linear correlation model for torrefied palm oil wastes. In *MATEC Web of Conferences, UTP-Ump Symposium on Energy Systems, Perak, Malaysia*. doi:10.1051/matecconf/201713104002.

Sadhukhan, J., Martinez-Hernandez, E., Murphy, R. J., Ng, D. K. S., Hassim, M. H., Ng, K. S., Kin, W. Y., Jaye, I. F., Hang, M. Y., Andiappan, V. 2018. Role of bioen-ergy, biorefinery and bioeconomy in sustainable development: Strategic path-ways for Malaysia. *Renewable and Sustainable Energy Reviews*, 81, 1966–1987.

Sarawak Energy, 2018. Palm oil mill effluent. www.sarawakenergy.com.my/index.php/r-d/biomass-energy/palm-oil-mill-effluent.

SEDA, 2018. Renewable energy policy in Malaysia. www.seda.gov.my/policies/national-renewable-energy-policy-and-action-plan-2009/.

Tan, R. R., Foo, D. C. Y. 2007. Pinch analysis approach to carbon-constrained energy sector planning. *Energy*, 32(8), 1422–1429.

Tan, R. R., Foo, D. C. Y. 2017. Carbon emissions pinch analysis for sustainable energy planning. *Encyclopedia of Sustainable Technologies*, 4, 231–237.

UN Framework Convention on Climate Change. 2015. Historic Paris agreement on climate change: 195 nations set path to keep temperature rise well below 2 degrees Celsius. https://unfccc.int/news/finale-cop21.

Yeomans, H., Grossmann, I. E., 1999. A systematic modeling framework of super-structure optimization in process synthesis. *Computers & Chemical Engineering*, 23(6), 709–731.

Zafar, S. 2018. Biomass wastes from palm oil mills. *BioEnergy Consult*. www.bioenergyconsult.com/palm-biomass/.

10

Carbon and Resource Constrained Energy Sector Planning for Taiwan

Jui-Yuan Lee

National Taipei University of Technology

Fossil fuels have been heavily exploited since the second Industrial Revolution began. The resulting carbon emissions are widely regarded as the main cause of global warming and climate change. Key mitigation technologies for reducing carbon emissions include CCS and renewables. According to a recent analysis of the International Energy Agency, renewables and CCS will contribute more than 50% of the cumulative emissions reductions by 2050. This chapter develops a mathematical programming model for optimal energy sector planning with CCS and renewables deployment. The model considers a variety of *carbon capture* (CC) options for the retrofit of individual thermal power generation units. For comprehensive planning, the *Integrated Environmental Control Model* (IECM) is employed to assess the performance and costs of different types of power generation units before and after CC retrofits. A case study of Taiwan's energy sector is presented to demonstrate the use of the model for complex decision-making and cost trade-offs in the deployment of CC technologies and additional low-carbon energy sources. Different scenarios are analyzed to identify the optimal strategy of energy mix to satisfy the electricity demand and various planning constraints.

10.1 Introduction

Climate change is largely due to the relentless rise in CO_2 levels in the atmosphere since the second Industrial Revolution, stemming from the world's heavy reliance on fossil fuels. Global electricity generation amounted to 24,973 TWh in 2016 (38.4% from coal, 3.7% from oil, and 23.2% from natural gas), while global CO_2 emissions from fuel combustion totaled 32,316 Mt (International Energy Agency (IEA), 2018). CO_2 emissions can be

137

reduced on both the supply and demand sides through CCS, increased use of renewable energy (RE) and energy efficiency enhancement. RE plays a central role in the transition to a low-carbon sustainable energy system; however, despite sharp cost reductions for solar photovoltaic (PV) and wind power, most renewables are still more expensive and less reliable than conventional fossil and nuclear energy in many parts of the world. On the other hand, CCS has been shown to be an integral part of any lowest-cost mitigation scenario with an increase in long-term global average temperature significantly less than 4°C (IEA, 2013). CCS allows significant emissions reductions in power generation and industrial processes, and provides the foundation for negative emissions. Without CCS, it is unlikely that the Paris Agreement commitments could be met (Haszeldine et al., 2018).

Carbon-constrained energy planning (CCEP) is an established area of research with an effort to reduce climate change effects. Several early CCEP techniques were developed under the framework of CEPA, which is an extension of conventional pinch analysis for heat and mass integration in the process industries to macro-scale applications such as regional and national electricity generation sectors (Tan and Foo, 2007). CEPA has been applied to electricity sectors in several countries such as Ireland (Crilly and Zhelev, 2008) and New Zealand (Atkins et al., 2010). Apart from CEPA approaches, mathematical programming models have also been developed for energy planning (Hashim et al., 2005; Muis et al., 2010). For further emissions reductions, CCS options have been included in CCEP (Tan et al., 2009; Pękala et al., 2010).

It is obvious from the literature that CCEP is an important approach for meeting energy demand and resource limits. Moreover, CCS has been recognized as a key technology to achieve the energy and environmental goals. In this chapter, a mathematical programming model is developed for optimal energy sector planning with deployment of CCS and renewables. The formulation considers a portfolio of different CCS options for each individual generation unit of fossil fuel power plants, and employs the IECM to assess the performance and costs of different types of generation units before and after CCS retrofit. Simulation results from IECM are then used as input data in the planning model for detailed and comprehensive analysis. In "Taiwan's Electricity Sector," "Plant Simulation," and "Other Planning Constraints" sections, a case study of Taiwan is described first, followed by the "Model Formulation" and "Results and Discussion" sections. Finally, the conclusions are given.

10.2 Taiwan's Electricity Sector

This case study considers medium-term planning of Taiwan's electricity sector for 2025, which is an initial checkpoint in Taiwan's energy transition. By the end of 2017, the total installed capacity in Taiwan was 49.8 GW;

the gross power generation in 2017 was 270.3 TWh (Bureau of Energy, 2019a). Thermal power plants using coal, oil, and natural gas accounted for 73.8% of the total installed capacity and 85.9% of the total power generation. Figure 10.1 shows the 2017–2031 load forecast provided by Taiwan Power Company (Taipower) (Taiwan Power Company, 2017). With the growing demand for electricity, the required power supply is expected to increase at an average annual rate of 1.4%. This implies a concurrent increase in carbon emissions in the business-as-usual (BAU) scenario. In response to climate change, a long-term national goal has been set for Taiwan to reduce greenhouse gas (GHG) emissions to no more than 50% of the 2005 level by 2050 (Environmental Protection Administration, 2019). The government also aims to meet an economy-wide target of reducing GHG emissions by 50% from the BAU level, or by 20% from the 2005 level, by 2030 (Executive Yuan, 2019). These targets are assumed in this case study to apply directly to the electricity sector. Using historical data (Bureau of Energy, 2019b; Taiwan Power Company, 2019), the sectoral carbon emissions in 2005 are calculated to be 97.3 Mt. The emissions limits for 2030 and 2050 are thus determined to be 77.8 and 48.6 Mt CO_2e, respectively. Assuming a quadratic emissions reduction trajectory from the 2017 level, as shown in Figure 10.2, the emissions limit for 2025 is estimated at 92 Mt CO_2e.

Table 10.1 shows the baseline energy mix in 2025, based on Taipower's long-term power development plan (Taiwan Power Company, 2016). The capacity factors are determined from Taipower's 2013–2017 statistical data (Taiwan Power Company, 2019) or taken from National Renewable Energy Laboratory estimates (NREL, 2019) (for small hydro, offshore wind, and geothermal). Given the controversy over nuclear power, two scenarios (with and without the use of the fourth nuclear power plant) are considered in this case study. Figure 10.3 shows that in both scenarios, thermal power plants using coal, oil, and natural gas comprise more than 75% of the total installed

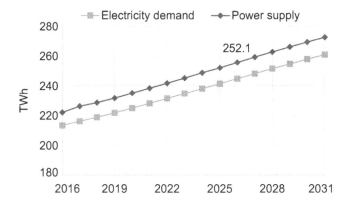

FIGURE 10.1
Forecast electricity demand and power supply for 2017–2031 (pumping energy excluded).

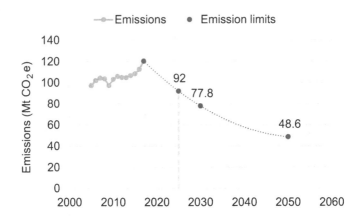

FIGURE 10.2
Carbon emissions and the reduction trajectory.

TABLE 10.1

Projected 2025 Baseline Generation Capacity

Source	Installed Capacity (GW)	Capacity Factor (%)
Coal (Taipower)	11.2	90.5
Coal (IPPs)	3.0971	78.6
Oil	0.3111	29.4
Natural gas (Taipower)	18.3546	50–70
Natural gas (IPPs)	4.61	44.6
Nuclear	2.7	90.1
Pumped hydro	2.602	14
Conventional hydro	2.0893	27.9
Small hydro	0.013	40.2
Onshore wind	0.8458	27.2
Offshore wind	0.74	39
PV	2.0436	13.3
Biomass, waste, and biogas	0.6235	46
Geothermal	0.05	85

IPPs, independent power producers.

capacity. This indicates the need for CCS and the expansion of RE in order for the electricity sector to meet the emissions limit (92 Mt CO_2e) and the required power supply of 252.1 TWh (see Figure 10.1) in 2025. Note that the load forecast in Figure 10.1 does not include pumping energy, which is about 2% of the power supply (Taiwan Power Company, 2019). Therefore, the overall energy demand to be satisfied in 2025 is 257.1 TWh.

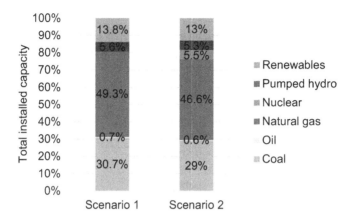

FIGURE 10.3
Baseline energy mix in 2025.

10.3 Plant Simulation

Table 10.2 lists the thermal power plants and generation units in the 2025 energy mix. Note that SPPs are small power plants on offshore islands, where all the generators are oil-fired (with installed capacities ranging from 0.3 to 11 MW) and are lumped into a single unit, SPPO. Similarly, independent power producers consist of nine private power plants, where coal-fired and gas-fired generation units are lumped into IPPC and IPPG, respectively. In this case study, only Taipower's large generation units (excluding SPPs and IPPs) are considered for CCS, except HTC3, HTC4, and HTG3-HTG5, which are due to be decommissioned in 2026–2027. The CC options are post-combustion technologies that are relatively mature and suitable for existing power plants in Taiwan, including amine, ammonia (NH_3), and membrane systems. Oxy-fuel combustion and chemical (calcium) looping systems are excluded because these technologies are more suited to new power plants than existing ones. For amine and ammonia systems, there is the option of adding an auxiliary gas boiler to generate low-pressure steam for sorbent regeneration. An auxiliary steam turbine can then be used in conjunction with the gas boiler to generate additional power and/or low-pressure steam. Table 10.3 shows the compatibility of the CC technologies with the thermal power plants. The membrane system requires a high concentration of CO_2 in the flue gas, and is therefore not suitable for gas-fired plants.

All candidate plants for CCS retrofit were simulated using IECM (Berkenpas et al., 2018) to assess their performance before and after the retrofit. First, plant type and technologies for combustion/post-combustion

TABLE 10.2

Thermal Power Plants and Generation Units in the 2025 Energy Mix

Plant	Unit	Installed Capacity (GW)	Fuel Type	For CCS?
HH	HH1	0.9	Natural gas	Yes
LK	LK1-LK3	0.8×3	Coal	Yes
SA	SA1	0.6	Coal	Yes
TT	TT1, TT2	0.7427×2	Natural gas	Yes
	TT3-TT6	0.7247×4	Natural gas	Yes
	TT7-TT10	0.792×4	Natural gas	Yes
GY	GY1	1.2	Natural gas	Yes
TS	TS1-TS3	0.8926×3	Natural gas	Yes
	TS4, TS5	0.9×2	Natural gas	Yes
	TS6	0.3212	Natural gas	Yes
TC	TC1-TC10	0.55×10	Coal	Yes
HTC	HTC3, HTC4	0.55×2	Coal	No
HTG	HTG1	0.9	Natural gas	Yes
	HTG3-HTG5	0.4452×3	Natural gas	No
NP	NP1-NP3	0.2888×3	Natural gas	Yes
	NP4	0.2514	Natural gas	Yes
TLC	TLC1, TLC2	0.8×2	Coal	Yes
TLG	TLG6	0.55	Natural gas	Yes
SPPs	SPPO	0.3111	Oil	No
IPPs	IPPC	3.0971	Coal	No
	IPPG	4.61	Natural gas	No

controls, as well as water and solids management were chosen according to actual equipment of the plant. Next, the capacity factor and ambient conditions (temperature, pressure, and humidity) were set for the overall plant. The installed capacity and turbine/boiler type were also entered, while the other parameters for the performance were calculated by default. Fuel properties (higher heating value and composition) and costs were then specified according to Taipower's purchase information. In addition, the ash properties of coal (oxide content) need to be specified too. The ash content determines the resistivity of ash, and hence the specific collection area of the cold-side electrostatic precipitator.

Simulation results of the power generation units before CCS were verified by comparing the IECM outputs with the actual performance. For the simulation of retrofitted power plants, the CC system is selected on the plant design screen and configured on the corresponding "config" screen (Berkenpas et al., 2018). It should be noted that, although not exactly accurate, cost estimates of IECM give correct relative costs of electricity, and are thus useful for planning Taiwan's energy sector in 2025. The parameters obtained from IECM simulations for the case study are given in Tables A1–A8 in the

TABLE 10.3

Compatibility between Thermal Power Plants and CC Technologies

	FG+	FG+-B	FG+-B-T	MEA	MEA-B	MEA-B-T	NH$_3$	NH$_3$-B	NH$_3$-B-T	PM
HH	O	O	O	O	O	O	O	O	O	X
LK	O	O	O	O	O	O	O	O	O	O
SA	O	O	O	O	O	O	O	O	O	O
TT	O	O	O	O	O	O	O	O	O	X
GY	O	O	O	O	O	O	O	O	O	X
TS	O	O	O	O	O	O	O	O	O	X
TC	O	O	O	O	O	O	O	O	O	O
HTC[a]	-	-	-	-	-	-	-	-	-	-
HTG	O	O	O	O	O	O	O	O	O	X
NP	O	O	O	O	O	O	O	O	O	X
TLC	O	O	O	O	O	O	O	O	O	O
TLG	O	O	O	O	O	O	O	O	O	X
SPPs[a]	-	-	-	-	-	-	-	-	-	-
IPPs[a]	-	-	-	-	-	-	-	-	-	-

FG+, CO_2 absorber using 30 wt% monoethanolamine (MEA) solvent with corrosion inhibitors; MEA, CO_2 absorber using 15–20 wt% MEA solvent without the inhibitor; NH$_3$, ammonia CO_2 scrubber; B, with an auxiliary gas boiler; T, with an auxiliary steam turbine; PM, polymer-based membrane.

[a] Not considered for CCS.

Appendix.[1] Note that the levelized costs of electricity from thermal power generation units are calculated based on costs in 2016. The parameters for low-carbon energy sources are given in Table A9.[1]

10.4 Other Planning Constraints

Apart from the energy demand and emissions limit discussed in Section 10.2, there are also limits on natural gas consumption and RE expansion. According to the Energy Transition White Paper (Ministry of Economic Affairs, 2018), the natural gas supply in Taiwan is expected to reach 32.7 Mt/y by 2025, of which about 80% can be used for power generation. Therefore, the gas consumption limit for 2025 is set to 26.16 Mt. Table 10.4 shows the government's current targets for RE development (Bureau of Energy, 2019c). The limit to the additional capacity of renewables is calculated by subtracting the baseline capacity in Table 10.1 from the target value.

---

[1] Note: available on the book support website.

TABLE 10.4

Government Targets for Renewables

Source	2025 Installed Capacity (GW)	Capacity Expansion Limit (GW)
Conventional hydro	2.089	0
Small hydro	0.061	0.048
Onshore wind	1.2	0.354
Offshore wind	3	2.26
PV	20	17.956
Biomass, waste, and biogas	0.813	0.189
Geothermal	0.2	0.15

10.5 Model Formulation

In this section, a mathematical model is developed for the Taiwan case study for optimal deployment of CC technologies and renewables under carbon and resource constraints. The model requires data for power generation, CC performance, and electricity cost. With these input data, the model is able to determine the strategy that minimizes the cost increase due to CC retrofits and new, low-carbon power plants, while satisfying the specified emission and resource limits. The formulation is presented below. Notation used is given in the Nomenclature.

Equations 10.1–10.4 deal with the selection of CC technologies for thermal power plants and generation units. Equation 10.1 states that only one CC technology can be selected for the power plant to be retrofitted. Equation 10.2 defines the forbidden ($M_{pk} = 0$) and allowable ($M_{pk} = 1$) matches considering the compatibility between specific power generation and CC technologies. Equation 10.3 ensures that the selection made for plant p applies to all its units $j \in J_p$. If none of the units in plant p is to be retrofitted using technology $k \left(\sum_{j \in J_p} y_{jk} = 0 \right)$, no such selection should be made at the plant level ($y_{pk} = 0$), as given in Equation 10.4.

$$\sum_{k \in \mathbf{K}} y_{pk} \leq 1 \quad \forall p \in \mathbf{P} \tag{10.1}$$

$$y_{pk} \leq M_{pk} \quad \forall k \in \mathbf{K}, p \in \mathbf{P} \tag{10.2}$$

$$y_{jk} \leq y_{pk} \quad \forall j \in \mathbf{J}_p, k \in \mathbf{K}, p \in \mathbf{P} \tag{10.3}$$

$$y_{pk} \leq \sum_{j \in J_p} y_{jk} \quad \forall k \in \mathbf{K}, p \in \mathbf{P} \tag{10.4}$$

The power generation of unit j after plant retrofits and the corresponding increase in electricity cost are given by Equations 10.5 and 10.6, respectively. Note that if unit j is not retrofitted $\left(\sum_{k \in \mathbf{K}} y_{jk} = 0 \right)$, the energy output will remain at the baseline level ($q_j = G_j CF_j TR_j$) with no additional cost ($x_j = 0$). The carbon emissions of unit j are calculated using Equation 10.7. Similarly, this equation gives the results of retrofit or the baseline carbon emission levels – if unit j is unmodified. It should also be noted that the carbon footprints and levelized costs of electricity in Equations 10.6 and 10.7 are based on gross generation.

$$q_j = CF_j T \left[G_j R_j \left(1 - \sum_{k \in \mathbf{K}} y_{jk} \right) + \sum_{k \in \mathbf{K}} G_{jk} R_{jk} y_{jk} \right] \quad \forall j \in \mathbf{J} \tag{10.5}$$

$$x_j = CF_j T \sum_{k \in \mathbf{K}} \left(G_{jk} LCOE_{jk} - G_j LCOE_j \right) y_{jk} \quad \forall j \in \mathbf{J}^R \tag{10.6}$$

$$e_j = CF_j T \left[G_j C_j \left(1 - \sum_{k \in \mathbf{K}} y_{jk} \right) + \sum_{k \in \mathbf{K}} G_{jk} C_{jk} y_{jk} \right] \quad \forall j \in \mathbf{J} \tag{10.7}$$

Equation 10.8 dictates that the energy demand should be satisfied. If the existing mix of energy sources cannot satisfy the demand, new power plants will be needed. Note that the last two terms on the left-hand side represent power generation from existing and additional low-carbon energy sources. Equation 10.9 imposes a carbon footprint constraint on power generation. In this constraint, the first term on the left-hand side gives the contribution of thermal power plants (after retrofit), while the next two terms give those of the existing and new, low-carbon power plants.

$$\sum_{j \in \mathbf{J}} q_j + T \left(\sum_{i \in \mathbf{I}^{ex}} G_i^{ex} CF_i R_i + \sum_{i \in \mathbf{I}^{new}} g_i^{new} CF_i R_i \right) \geq D \tag{10.8}$$

$$\sum_{j \in \mathbf{J}} e_j + T \left(\sum_{i \in \mathbf{I}^{ex}} G_i^{ex} CF_i C_i + \sum_{i \in \mathbf{I}^{new}} g_i^{new} CF_i C_i \right) \leq E^{lim} \tag{10.9}$$

Equation 10.10 imposes a gas consumption constraint on the thermal power plants.

$$\sum_{j \in \mathbf{J}} u_j^{\text{tot}} \leq U^{\text{lim}} \tag{10.10}$$

where the gas consumption of unit j is given by Equation 10.11.

$$u_j^{\text{tot}} = CF_j T \left[G_j U_j \left(1 - \sum_{k \in \mathbf{K}} y_{jk} \right) + \sum_{k \in \mathbf{K}} G_{jk} U_{jk} y_{jk} \right] \quad \forall j \in \mathbf{J} \tag{10.11}$$

The capacity constraint on RE expansion is given in Equation 10.12.

$$g_i^{\text{new}} \leq G_i^{\text{lim}} \ \forall i \in \mathbf{I}^{\text{new}} \tag{10.12}$$

The objective function is to minimize the total cost increase, which consists of the additional cost incurred by CC retrofits and the cost of power generation using new, low-carbon energy sources, as given in Equation 10.13.

$$\min z = \sum_{j \in \mathbf{J}^R} x_j + T \sum_{i \in \mathbf{I}^{\text{new}}} g_i^{\text{new}} CF_i LCOE_i \tag{10.13}$$

Equations 10.1–10.13 constitute a mixed integer linear programming (MILP) model, for which global optimality can be guaranteed without major computational difficulties. The model is implemented and solved in GAMS (Rosenthal, 2018), utilizing CPLEX as the MILP solver. Solutions were found with negligible processing time for both scenarios.

10.6 Results and Discussion

Two scenarios are considered in this case study to examine the role of nuclear power in the energy mix. In both scenarios, the effect of increasing the capacity factor of Taipower's gas-fired plants (from 50% to 60% and to 70%) is further analyzed. This represents a recent change of the gas-fired power plants from being load-following plants to being more like baseload plants. Table 10.5 shows the baseline installed capacity, power generation, and carbon emissions for the individual cases. It can be seen that in both scenarios, when the gas-plant capacity factor is low (i.e., 50%), the baseline energy mix is incapable of meeting the energy demand of 257.1 TWh. At the same time, the baseline emissions exceed the emissions limit of 92 Mt in both scenarios. Therefore, not only will CCS be necessary but also additional power generation from renewables or auxiliary steam turbines may be required. The objective of the planning is to identify the optimal strategy that minimizes the additional cost to the electricity sector.

TABLE 10.5

Baseline Energy Mix Conditions

Parameter	Scenario 1: Without Nuclear Energy			Scenario 2: With Nuclear Energy		
Capacity factor of Taipower gas plants	50%	60%	70%	50%	60%	70%
Baseline installed capacity (GW)	46.58	46.58	46.58	49.28	49.28	49.28
Baseline power generation (TWh)	221.326	237.097	252.868	242.637	258.408	274.178
Baseline carbon emissions (Mt)	129.748	135.779	141.809	129.748	135.779	141.809
Baseline gas consumption (Mt)	13.738	15.931	18.124	13.738	15.931	18.124

Solving the MILP model for this case study (Equation 10.13 subject to Equations 10.1–10.12) with the input data in Tables 10.1–10.4 and A1–A9 yields the results in Table 10.6. The model involves 822 constraints and 811 variables (including 553 binaries), and is solved in 0.5 CPU seconds. Overall, the increase in electricity cost (from the baseline) becomes smaller as the gas-plant capacity factor increases, as shown in Figure 10.4, and the use of nuclear energy reduces the additional costs. In scenario 1, increasing the power output from the gas plants reduces the need for CCS and additional power generation capacity. In other words, less capacity of thermal power plants is retrofitted, fewer auxiliary steam turbines are used, and a lower amount of RE is required. The same situation is observed in scenario 2 when the gas plants operate at capacity factors of 50% and 60%. However, the

TABLE 10.6

Results Summary for the Case Study

Parameter	Scenario 1: Without Nuclear Energy			Scenario 2: With Nuclear Energy		
Capacity factor of Taipower gas plants	50%	60%	70%	50%	60%	70%
Electricity cost increase (billion USD)	6.805	4.956	3.245	3.870	2.154	1.279
Final installed capacity (GW)	54.222	51.556	49.113	53.253	50.973	49.28
Coal	4.197	4.797	4.797	6.447	6.447	9.747
Coal-CCS	10.1	9.5	9.5	7.85	7.85	4.55
Gas	22.965	22.713	22.965	22.24	22.965	10.311
Gas-CCS	0	0.251	0	0.725	0	12.654
Auxiliary steam turbine	3.993	2.32	0.873	1.624	0	0
Oil	0.311	0.311	0.311	0.311	0.311	0.311
Nuclear	0	0	0	2.7	2.7	2.7
Pumped hydro	2.602	2.602	2.602	2.602	2.602	2.602
Renewables	10.055	9.061	8.065	8.754	8.099	6.405
Final power generation (TWh)	257.1	257.1	257.1	257.1	257.1	257.155
Final carbon emissions (Mt)	90.964	91.86	91.847	91.951	91.684	91.85
Final gas consumption (Mt)	20.987	20.147	19.707	16.689	15.931	18.124

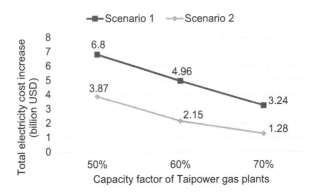

FIGURE 10.4
Total cost increase versus the gas plant capacity factor.

capacity of thermal power plants for CCS is significantly increased when the gas-plant capacity factor increases to 70% (see the last column of Table 10.6), although there is no need for auxiliary steam turbines or additional RE. Table 10.7 shows the retrofitted power generation units and the selected CC technologies in each case. It can be seen that coal-fired generation units are prime targets for CCS, except in the last case, where most of the retrofitted units are gas-fired. In addition, the FG+ amine system is the most selected CC option, because it has higher energy output ratios and is less expensive for gas-fired units. Figures 10.5a and b shows the additional RE capacities for scenarios 1 and 2, respectively. It appears that the sources with lower costs or higher capacity factors (i.e., small hydro, geothermal, bioenergy, and onshore

TABLE 10.7

CCS Retrofits for the Case Study

Scenario	Taipower Gas-Plant Capacity Factor	Unit	CC Technology
1: Without nuclear energy	50%	LK1-LK3, SA1, TC1-TC10, TLC1, TLC2	FG+-B-T
	60%	LK1-LK3, NP4, TLC1, TLC2	FG+
		TC1-TC10	FG+-B-T
	70%	TC1-TC10, TLC1, TLC2	FG+
		LK1-LK3	FG+-B-T
2: With nuclear energy	50%	LK1-LK3, TT6, TLC1, TLC2	FG+
		TC4-TC10	FG+-B-T
	60%	LK1-LK3, TC4-TC10, TLC1, TLC2	FG+
	70%	HH1, TT1-TT7, GY1, TS1-TS5, HTG1	FG+
		LK1-LK3, TLC1, TLC2	NH$_3$
		TC10	PM

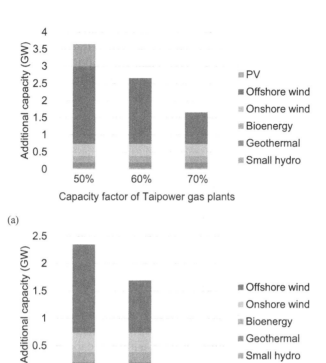

(a)

(b)

FIGURE 10.5
Expansion of renewables for (a) scenario 1 and (b) scenario 2.

wind) are preferred, while additional PV is only required when all the others have reached their capacity limits. However, offshore wind has the largest share (greater than 55%) of the RE expansion.

In addition, the results demonstrate the role of nuclear power as an economical low-carbon option in energy transition. The use of nuclear energy reduces the need for additional power generation capacity from renewables in 2025, thus relieving the pressure to achieve ambitious RE targets in the medium term. This might seem to delay the deployment of renewables, but is actually to buy time for RE technologies to be deployed progressively in the long term, without major impacts on grid stability and electricity cost. With similar effects, increasing the capacity factor of load-following (i.e., gas-fired) power plants reduces the operating reserve, and hence grid flexibility, while causing difficulties in maintenance scheduling. Therefore, the preferred strategy would be to use nuclear energy with gas-fired plants operating at optimized capacity factors. It should be noted that there is no direct comparison between the results for scenarios 1 and 2, because of different baseline conditions.

However, the total levelized annual cost of nuclear power generation can be estimated to be 0.68 billion USD, which is less than the difference in the resulting electricity cost increase between scenarios 1 and 2 (see Figure 10.4). This indicates the economic benefit of nuclear energy in this specific case.

10.7 Conclusion

A mathematical model for optimal energy sector planning with CCS and renewables deployment has been developed in this chapter. The MILP model considers the energy demand to be met under carbon footprint, resource, and capacity constraints, with the objective of determining the optimal strategy to minimize the total electricity cost increase from CC retrofits and additional low-carbon energy sources. In addition, the model allows the selection of CC technologies for individual power generation units with compatibility considerations, based on the performance and cost parameters obtained through IECM simulation. A case study of Taiwan's electricity sector planning was solved to illustrate the proposed approach. Two scenarios for the 2025 energy mix and the effect of increasing the capacity factor of gas-fired power plants were analyzed. The results show that using nuclear power (in scenario 2) reduces the additional costs to the sector. Such a model can thus be used to gain insights into regional, sectoral, or national energy planning and provide decision support for policy makers.

Nomenclature

Indices and Sets

$i \in \mathbf{I}$	low-carbon energy sources
$i \in \mathbf{I}^{\text{ex}}$	existing low-carbon energy sources
$i \in \mathbf{I}^{\text{new}}$	additional low-carbon energy sources
$j \in \mathbf{J}$	thermal power generation units
$j \in \mathbf{J}_p$	thermal power generation units in plant p
$j \in \mathbf{J}^{\text{R}}$	thermal power generation units to be retrofitted
$k \in \mathbf{K}$	CC technologies
$p \in \mathbf{P}$	thermal power plants

Parameters

C_i	carbon footprint of energy source i (kg CO_2e/kWh)
C_j	carbon footprint of generation unit j (kg CO_2e/kWh)

C_{jk}	carbon footprint of generation unit j after retrofitting with CC technology k (kg CO_2e/kWh)
CF_i	capacity factor of energy source i
CF_j	capacity factor of generation unit j
D	energy demand (GWh)
E^{lim}	carbon emissions limit (kt CO_2e)
G_i^{ex}	installed capacity of existing energy source i (GW)
G_i^{lim}	capacity limit for energy source i (GW)
G_j	installed capacity of generation unit j (GW)
G_{jk}	installed capacity of generation unit j after retrofitting with CC technology k (GW)
$LCOE_i$	levelized cost of electricity from energy source i (USD/kWh)
$LCOE_j$	levelized cost of electricity from generation unit j (USD/kWh)
$LCOE_{jk}$	levelised cost of electricity from generation unit j after retrofitting with CC technology k (USD/kWh)
M_{pk}	binary indicating the compatibility of CC technology k with plant p
R_i	energy output ratio of source i
R_j	energy output ratio of generation unit j
R_{jk}	energy output ratio of generation unit j after retrofitting with CC technology k
T	annual operating time (h)
U^{lim}	gas consumption limit (kt)
U_j	gas consumption of generation unit j (kg/kWh)
U_{jk}	gas consumption of generation unit j after retrofitting with CC technology k (kg/kWh)

Variables

e_j	final carbon emissions from generation unit j (kt CO_2e)
g_i^{new}	installed capacity of additional energy source i (GW)
q_j	final power output from generation unit j (GWh)
u_j^{tot}	total gas consumption of generation unit j (kt)
x_j	additional cost of retrofitting generation unit j (million USD)
y_{jk}	binary indicating the use of CC technology k for retrofitting generation unit j
y_{pk}	binary indicating the selection of CC technology k for plant p

References

Atkins, M. J., Morrison, A. S., Walmsley, M. R. W. 2010. Carbon Emissions Pinch Analysis (CEPA) for emissions reduction in the New Zealand electricity sector. *Applied Energy*, 87, 982–987.

Berkenpas, M. B., Fry, J. J., Kietzke, K., Rubin, E. S. 2018. *IECM User Documentation: User Manual*. The Integrated Environmental Control Model Team, Carnegie Mellon University, Pittsburgh, PA.

Bureau of Energy, 2019a. Energy statistical annual reports. www.moeaboe.gov.tw/ECW/english/content/ContentLink.aspx?menu_id=1540.

Bureau of Energy, 2019b. Greenhouse gases. www.moeaboe.gov.tw/ecw/populace/content/SubMenu.aspx?menu_id=114.

Bureau of Energy, 2019c. Policy & program. www.moeaboe.gov.tw/ECW/populace/content/SubMenu.aspx?menu_id=48.

Crilly, D., Zhelev, T., 2008. Emissions targeting and planning: An application of CO_2 emissions pinch analysis (CEPA) to the Irish electricity generation sector. *Energy*, 33, 1498–1507.

Environmental Protection Administration, 2019. Greenhouse gas reduction and management act. https://ghgrule.epa.gov.tw/greenhouse/greenhouse_page/23.

Executive Yuan, 2019. Government agency news. www.ey.gov.tw/Page/AE5575EAA0A37D70/381b1f63-4a6f-4fd3-8bd3-5ecb2bdb9d3c.

Hashim, H., Douglas, P., Elkamel, A., Croiset, E., 2005. Optimization model for energy planning with CO_2 emission considerations. *Industrial & Engineering Chemistry Research*, 44, 879–890.

Haszeldine, R. S., Flude, S., Johnson, G., Scott, V., 2018. Negative emissions technologies and carbon capture and storage to achieve the Paris Agreement commitments. *Philosophical Transactions of the Royal Society A: Mathematical, Physical and Engineering Sciences*, 376.

International Energy Agency (IEA), 2018. *Key World Energy Statistics*. IEA, Paris.

IEA, 2013. *Technology Roadmap: Carbon Capture and Storage*. OECD/IEA, Paris.

Muis, Z. A., Hashim, H., Manan, Z. A., Taha, F. M., Douglas, P. L. 2010. Optimal planning of renewable energy-integrated electricity generation schemes with CO_2 reduction target. *Renewable Energy*, 35, 2562–2570.

Ministry of Economic Affairs, 2018. *Energy Transition White Paper. Ministry of Economic Affairs*, Taipei, Taiwan.

NREL, 2019. Utility-scale energy technology capacity factors. www.nrel.gov/analysis/tech-cap-factor.html.

Pękala, Ł. M., Tan, R. R., Foo, D. C. Y., Jeżowski, J. M. 2010. Optimal energy planning models with carbon footprint constraints. *Applied Energy*, 87, 1903–1910.

Rosenthal, R. E. 2018. *GAMS – A User's Guide*. GAMS Development Corporation, Washington, DC.

Taiwan Power Company, 2016. *Long-Term Power Development Plan*. Taiwan Power Company, Taipei, Taiwan.

Taiwan Power Company, 2017. *Long-Term Load Forecast*. Taiwan Power Company, Taipei, Taiwan.

Taiwan Power Company, 2019. Statistical data. www.taipower.com.tw/tc/page.aspx?mid=43&cid=29&cchk=34db42ba-62b1-4684-9fc8-59881779ac23.

Tan, R. R., Foo, D. C. Y. 2007. Pinch analysis approach to carbon-constrained energy sector planning. *Energy*, 32, 1422–1429.

Tan, R. R., Ng, D. K. S., Foo, D. C. Y. 2009. Pinch analysis approach to carbon-constrained planning for sustainable power generation. *Journal of Cleaner Production*, 17, 940–944.

Index

Printed in the United States
by Baker & Taylor Publisher Services